ELEMENTARY
DIFFERENTIAL AND INTEGRAL
CALCULUS

CAMBRIDGE
UNIVERSITY PRESS

University Printing House, Cambridge CB2 8BS, United Kingdom

Cambridge University Press is part of the University of Cambridge.

It furthers the University's mission by disseminating knowledge in the pursuit of education, learning and research at the highest international levels of excellence.

www.cambridge.org
Information on this title: www.cambridge.org/9781316612699

© Cambridge University Press 1936

Elementary Differential Calculus
First edition 1927
Second edition 1934

Elementray Integral Calculus
First edition 1926

First issues in this form 1936
First paperback edition 2016

A catalogue record for this publication is available from the British Library

ISBN 978-1-316-61269-9 Paperback

ELEMENTARY DIFFERENTIAL AND INTEGRAL CALCULUS

by

G. LEWINGDON PARSONS, M.A.

Chief Mathematical Master at Merchant Taylors' School

CAMBRIDGE

AT THE UNIVERSITY PRESS

1936

PART I

ELEMENTARY
DIFFERENTIAL CALCULUS

PREFACE

THE present volume is intended for the general purpose of preparing candidates for the Higher Certificate Examinations. Like its companion volume, it owes its origin to lecture notes given in preparation for this examination, and can thus be said to have stood, in part at any rate, the test of experience. The author feels that in Differential Calculus (just as in Integral Calculus) there is too wide a gap between the easy introductions to the Calculus, which do not go far enough, and the advanced text-books, in which the main principles are in danger of being lost in a mass of subsidiary detail. To the mathematical specialist these details are important and often essential, but the author is strongly of opinion that even the would-be mathematical specialist will acquire mathematical precision much more easily if he has previously worked through a book such as this, by way of general introduction to the subject. To those, on the other hand, who wish to regard the Calculus as a working tool and not as a subject for exhaustive enquiry, the finer points are unnecessary, and it may be hoped that they, too, will find most of their needs supplied in the present volume.

A few applications to Algebra, Mechanics, etc., are given; these simple methods are given in so few elementary text-books that it seemed worth while to include them. The order of the different parts of the subject is a matter of taste, and some of the examples may well be omitted by all but the most intelligent pupils. A short historical sketch has been added. Much of the history of the invention of the Calculus is obscure, and much more is controversial, but the author feels strongly that some attempt ought to be made to teach it.

The Oxford and Cambridge Joint Board have again kindly granted permission to use questions set in their Higher Certificate Examinations, and such examples are denoted in the text by an asterisk.

In conclusion the author's sincere thanks are due to the publishers, whose accurate work and skilful cooperation have greatly facilitated the labour of preparing the book for the press.

G. L. P.

August, 1927.

A number of corrections have been made in the second edition. An appendix on Newton's Approximation and a set of Miscellaneous Examples have been added.

G. L. P.

November, 1933

CONTENTS

CHAPTER IV

MAXIMA AND MINIMA

CHAPTER V

APPLICATION TO GEOMETRY OF CURVES

CHAPTER VI

THE EXPANSION OF A FUNCTION AS AN INFINITE SERIES

CHAPTER VII

MISCELLANEOUS APPLICATIONS TO ALGEBRA, TRIGONOMETRY AND MECHANICS

CHAPTER VIII

CURVATURE OF PLANE CURVES

CHAPTER IX

ENVELOPES AND ASSOCIATED LOCI

CHAPTER X
SIMPLE CURVE TRACING

CHAPTER XI
RECTILINEAR ASYMPTOTES

CHAPTER XII
PARTIAL DIFFERENTIATION

HISTORICAL SKETCH

THE history of the Differential Calculus is largely concerned with the thoughts and writings of two men, Newton and Leibnitz. There are, however, two others who may justly be regarded as their precursors, and who, in a sense, may be said to have had some part in the invention of the Differential Calculus.

The first of these was Pierre de Fermat, who was born in 1601. In public life he was a student of law and a member of the French Parliament, but he devoted his spare time to the study of mathematics and succeeded in leaving his mark on almost every branch of the subject as it was then constituted. In 1628 or 1629 he published a method for finding maxima and minima which consisted in the simultaneous solution of the equations $f(x) = 0$, $f(x + e) = 0$, where e is a small but finite quantity. We can see at once, by using Taylor's Theorem, that this method and our own are virtually the same. He was enabled to solve many geometrical questions in this way, although some controversy was aroused, in which Descartes and Desargues objected to the method, while Pascal and Roberval supported it. Unfortunately Fermat was of a retiring and secretive disposition and often gave his results without indicating how they had been obtained. The eminent French mathematicians, Lagrange, Laplace and Fourier endeavoured (with a laudable patriotism) to attribute the invention of the Calculus to Fermat, but as another distinguished Frenchman, Poisson, has pointed out, the term Differential Calculus can only logically be

applied to a series of rules for finding differential coefficients, and not to any one application or series of
applications. Judged by this standard the claim to regard
Fermat as the originator of the Calculus fails completely.

The second "forerunner" of the Calculus was Isaac
Barrow (1630–77). By his method of the "Differential
Triangle" published in 1670, he virtually gave the rule
we should denote by $\dfrac{dy}{dx} = \tan \psi$. To Barrow's credit stands
one of the most generous acts in the whole history of
mathematics, namely, that in 1669 he resigned his chair
of mathematics at Cambridge in favour of one of his own
pupils, Sir Isaac Newton.

We now come to the two main figures in this epoch-
making discovery. Isaac Newton, considered by many to
be the greatest genius produced by the human race, was
born in 1642. He went to Cambridge in 1660 and quickly
mastered all the mathematics then known. His discovery
of the Binomial Theorem before he had reached the age
of 23 or taken his B.A. degree is but a single instance
of the power of his intellect. In the same year (1665) he
invented a method of dealing with varying quantities,
which he called the Method of Fluxions. This can best
be explained in Newton's own words: "in the equation
$y = x^2$, if y represent the length of the space at any time
described, which (time) another space x, by increasing with
uniform celerity $\dot{x}$, measures and exhibits as described,
then $2x\dot{x}$ will represent the celerity with which the space y
is described at the same moment and contrariwise." In our
present notation we see that the result would be expressed

$$\frac{dy}{dt} = 2x\,\frac{dx}{dt}.$$

The quantity $\dot{x}$ was called the fluxion of x and x was called the fluent. By expanding any algebraic functions by the aid of his Binomial Theorem, Newton was able to write any expression of which he desired to find the fluxion as an infinite series and so to perform the process required. In the same way the "fluents" of many forms were found by him.

The year of this discovery was that of the Great Plague, during which Newton retired from Cambridge to his home in Lincolnshire. Here he seems to have spent his time in perfecting the method and in investigating various applications with regard to tangents, curvature, and the theory of equations. Other matters occupied him on his return to Cambridge, but in 1671 he wrote a treatise on the subject, called "The Method of Fluxions," but owing to a natural modesty and an almost fanatical shrinking from controversy, which seems to have been one of his marked characteristics, he refused to publish it. It was not in fact published until 1736, nine years after his death; but there is no doubt as to when it was written, and many of Newton's friends were in possession of his results some time before this. This is one of the most unfortunate circumstances in the whole history of mathematics. Had Newton published his results in 1671, the bitter controversy between the adherents of Newton and those of Leibnitz, which finally caused English mathematicians to isolate themselves for over 100 years, could never have arisen. For, as a matter of fact, in 1671 Leibnitz had not commenced the study of mathematics at all.

We are not, in this sketch, concerned with the numerous contributions of Newton to other branches of mathematical knowledge, but we cannot pass on without mentioning his great work, the *Principia* (1687), in which he systematized the whole treatment of mechanics, both terrestrial and celestial, and which has remained, until quite recent times, the basis of our knowledge of the laws of the universe. In this, however, he used geometry and not fluxions.

The other great name in the history of the invention of the Calculus is that of G. W. Leibnitz (or Leibniz) (1646–1716). He was born in Leipzig but did not take up the study of mathematics until 1672, when he made the acquaintance of C. Huygens (Huyghens) in Paris. He visited London in 1673 and returned to Paris, which he left in 1676. Before leaving Paris he wrote, but did not publish, a treatise on quadratures (i.e. the Integral Calculus). On October 29, 1675, we first find in his notes the notation of the Calculus as we know it to-day. At first he wrote $\frac{x}{d}$ (having noticed that differentiation reduced the degree of the expression), but he soon changed to dx, which quickly established itself as the simplest notation. He also introduced the sign of integration. At first he was greatly puzzled—and even fell into error—on the differentiation of products and quotients—a fact which may afford some consolation to beginners of to-day. But on November 21, 1675, he discovered the rule

$$y\,dx = d\,(xy) - x\,dy,$$

and also succeeded in solving a differential equation. In a paper of 1677 he gave correct results for quotients, products, powers and roots. He also invented, a little later, the theory of envelopes.

We proceed to give a short account of the rise of the famous controversy as to the invention of the Calculus.

In 1676 Leibnitz was in correspondence with Oldenburg, the secretary of the Royal Society, who mentioned that Newton had some important results. In reply to Leibnitz' request for further information, Newton wrote two letters to Oldenburg in which a variety of theorems were given, but the method of Fluxions was only referred to by means of two anagrams, from which it is difficult to imagine that Leibnitz derived much information. Leibnitz wrote a reply, dated June 21, 1677, in which he gave his rules for dy, dx and their use. Newton, in the first edition of the *Principia*, mentioned the fact that Leibnitz had discovered independently a method "which hardly differed from mine, except in his forms of words and symbols." The whole matter appeared to be settled quite satisfactorily, and so it rested for 20 years.

But in 1699 an insignificant mathematician called de Duillier hinted in a paper to the Royal Society that Leibnitz had derived his method from Newton's notes, to which he was supposed to have had access during his visit to London in 1673. Leibnitz entered a spirited protest, but was apparently pacified when the Royal Society stated that they "did not approve" the accusation. But the poison was at work. In 1704 Leibnitz published an anonymous review of Newton's recently completed *Optics and Quadrature of Curves*, and in this review there occurs a passage which many have taken to infer that Newton borrowed from Leibnitz. The latter afterwards denied that this was intended, but the weight of evidence is rather against him, especially as we know that he was not above an occasional forgery in the matter of dates, etc. It is quite probable that he wrote it in a spirit of revenge for de Duillier's insult of five years before.

John Keill, Professor of Physics at Oxford, in 1708 repeated de Duillier's charge, and, at Leibnitz' request, the Royal Society undertook to arbitrate in the matter. Their decision took the form of a most unsatisfactory document called "Commercium Epistolicum" (1712), which in tone was hostile and somewhat unfair to Leibnitz and failed to settle the matter under dispute. Thus originated a dispute which had most unfortunate results.

The English and Continental schools separated sharply, and for over 100 years English mathematicians ignored the rapid developments of analysis produced by the Continental school. Challenge problems were issued by first one side and then the other, and unfortunately in one of these Keill proposed to Bernoulli a question which he himself was unable to solve, a mistake whereby the prestige of the English mathematicians was considerably lowered. It was not till 1812 that Peacock, Babbage and Herschel succeeded in overcoming the prejudice of years and adopting the Continental notation.

Modern judgments of this dispute are not totally in agreement. It has been proved that Leibnitz did have access to some of Newton's papers, since notes are found in his work of results attributed by him to Newton. But what papers he saw we do not finally know. Many of the authorities doubt whether he saw enough of Newton's work on Fluxions to give him an indication of the method, as his own results were developed along quite different lines.

In any case he did not plagiarize in the matter of notation, and for that we must be profoundly grateful. His instinct for mathematical form led him to express his results in finite form, in contrast to Newton, who always used infinite series. It would seem, therefore, that he has the better right to be considered as the inventor of the Differential Calculus, as we understand it.

Of the immediate successors of Newton only two made any large contribution. Brook Taylor (1685–1731) stated the theorem known by his name, but his proof of it is worthless, as it ignores the convergency of the series. He also gives the method of changing the variable which we write

$$\frac{d^2 x}{dz^2} = -\frac{\dfrac{d^2 z}{dx^2}}{\left(\dfrac{dz}{dx}\right)^3},$$

and similar rules for higher derivatives.

Colin Maclaurin (1698–1746) wrote a treatise on Fluxions in 1742. In this he gave for the first time the correct discrimination of maxima and minima, also rules for multiple points. He was the inventor of pedal curves, and the originator of the theorem called by his name, though his proof is open to the same objection as Taylor's. After this little was done by English mathematicians to improve the Differential Calculus. They followed largely the ideas of Newton, who belonged to a geometrical school of thought rather than to an analytical one. As a result of this retrospective attitude, the next century was productive of very little mathematics of an original character in England.

On the Continent, however, the new doctrines flourished vigorously. Leibnitz found able allies in the celebrated Bernoulli family, nine members of which were noted mathematicians. James Bernoulli, who was Professor of Mathematics at Basel from 1687 to 1705, gave in 1694 the expressions for the radius of curvature in Cartesian and polar coordinates. He was the first to use the term "integral" calculus and to study the curve known as

"Bernoulli's Lemniscate." John (1667–1748) succeeded his brother at Basel in 1705 and did a great deal of work in Integral Calculus, the exponential theorem, and caustic curves. Nicholas, Daniel, the younger John, and the younger Nicholas were all celebrated exponents of the new methods, and the last of these proved in 1742 that

$$\frac{\partial^2 u}{\partial x \partial y} = \frac{\partial^2 u}{\partial y \partial x}.$$

G. F. L'Hospital, a pupil of Leibnitz, published in 1696 the first text-book of the subject, which was of great assistance in disseminating the new methods. Clairaut (1713–65), D'Alembert (1717–83), Lambert (1728–77) were responsible for many of the standard theorems in Differential Equations, Dynamics and Astronomy. J. P. de Gua (1713–86) was an opponent of the Calculus, and devoted his energy mainly to showing that problems relating to curves could be solved just as easily by Cartesian methods. We owe to him the "Analytical Triangle" which forms the basis of the "graphic reduction" mentioned in Ch. IX.

L. Euler (1707–83), whose name is so frequently found in mathematics, extended and commented on the work already done, and produced a quantity of original research in connection with the Calculus, Differential Geometry, and Infinite Series. On the latter subject he is, like many of his contemporaries, rather haphazard, and, though he warned his readers to take careful note of the convergency of their series, he himself often fell into grave error in this respect. His influence on the mathematics of the day was profound and he certainly deserves mention as one of the great men connected with the history of the Calculus.

Lagrange (1736–1813), Laplace (1749–1827), Legendre (1752–1813) were a trio of brilliant contemporaries who made many contributions to Analytical Mechanics and allied subjects arising out of the Calculus, but their work is, in the main, too advanced to be regarded as part of its elementary side.

The attitude of the inventors of the Differential Calculus towards its fundamental concepts was extremely vague. They had some misgivings on the subject of infinitely small quantities, but without them were unable to explain satisfactorily the processes employed. Many criticisms were levelled at the new methods on this score, amongst which those of Bishop George Berkeley are the best known. The attitude of the inventors can be aptly summed up in the words of D'Alembert's advice to doubting students, "Allez en avant; et la foi vous viendra." The validity of the method was defended solely by the fact that it reached no obviously incorrect results. It was left to the brilliant mathematicians of the new century, prominent amongst whom were Gauss (1777–1855), Abel (1802–29), and Cauchy (1789–1857) to found a new and rigid school of thought, and inaugurate a new period of mathematical precision which is still with us. They completely revised the principles of the Calculus, which they placed on a sure foundation by means of the Method of Limits, and thoroughly investigated the problems of convergency and continuity. It must not be thought that England had no part in this movement. Space prevents more than a passing mention of De Morgan, Boole, Cayley, Sylvester and others. Much could be written of this interesting period in the history of mathematics, and for the specialized student there is much stimulation and encouragement to be derived from the works of these men. But in view of the elementary character of this book it is not necessary to give any further details.

CHAPTER I

PRELIMINARY IDEAS

§ 1. Introduction: the Continuous Variable.

Before endeavouring to explain the processes of the Differential Calculus, we must first examine a few of the fundamental ideas which will continually be needed in this subject.

The reader is, no doubt, already familiar with the idea of a "variable" or "varying quantity." In dealing with elementary graphs, for instance, the idea must have occurred before. In the Calculus, however, we have to make one further restriction; namely, that the variable shall be a "continuous" variable. If x is the variable under consideration and if a and b are two "boundary values" of x (i.e. if only values of x lying between a and b are considered), then x must pass in succession through every conceivable value lying between a and b, irrational values included. In all that follows we shall assume that our variables are continuous.

There is a useful convention by which we denote variables by letters near the end of the alphabet, and also by the Greek equivalents of these letters, as, $u, v, w, x, y, z, \xi, \eta, \zeta$, and so on; reserving the letters at the beginning of both alphabets for constants.

§ 2. Functions.

Many of the problems of Mathematics are concerned with the dependence of one magnitude on another; or on the inter-dependence of two magnitudes. When this dependence is such that, given one of the magnitudes, we can always determine the other, then the second of these magnitudes is called a *function* of the first. Many examples of functions will at once occur to the reader. If for example the radius of a circle is known, then its area can at once be found. Thus the area of a circle is a function of its radius. Or again we can have a function of two or more variables, e.g. the volume of a cylinder is a joint function of the height and base-radius.

This definition of a function can usually be replaced by two simpler statements. These are

(1) If y is a function of a continuous variable x, then y and x must be connected by some equation;

and (2) The relationship between a function y and its variable (or, as it is sometimes styled, its argument) can be represented by a graph.

The proof of these two statements lies somewhat beyond the scope of this work, though on general grounds they are seen to be true. We shall therefore assume, in general, that the functions with which we deal can be represented graphically and also by an equation, for all except special values of the variables.

§ 3. Implicit and Explicit Functions.

There are, generally speaking, two types of equation that may occur as the definition of functions. These are (*a*) explicit equations, (*b*) implicit equations.

An explicit equation is one in which y is given directly in terms of x, and y is then called an *explicit function of x.*

The following are examples of equations defining *explicit functions*:

(1) $y = x^2 + x + 1$.

(2) $y = \sin x$.

(3) $y = \pm \sqrt{1 - x^2}$.

(4) $y = -\dfrac{2x + 1}{x + 1}$.

An implicit equation is one in which neither variable is given explicitly in terms of the other; but an equation connecting the two variables is given instead.

The following are equations defining *implicit functions*:

(5) $x^2 + y^2 = 1$.

(6) $xy + 2x + y + 1 = 0$.

(7) $x^2 + 2xy - y^2 - 1 = 0$.

The reader may notice that (3) and (5), also (4) and (6), are identical. In fact, almost any implicit equation can theoretically be converted into an explicit one, but the labour of performing the conversion is often prohibitive and the result too cumbersome to use, as can be seen by solving (7) for y in terms of x.

§ 4. Many-valued Functions.

One further point may be noticed. The implicit equation (5) when converted to the explicit form (3) gives rise to an ambiguous sign. In such cases y is spoken of as being a "double-valued" function of x, there being two distinct values of y to every value of x. [The fact that these values are equal in magnitude is incidental and is not necessarily always the case.] If the graphical mode of representation is adopted it will be found that the graph of such a function is a curve consisting of two branches.

The "inversion" of a single-valued function often gives rise to a many-valued function. For example, take the function $\sin x$. Now $y = \sin x$ and $x = \sin^{-1} y$ are on the face of it the same equation. But one value of y corresponds to every value of x, whereas in the second any number of values of x appear to correspond to a value of y. In such cases we usually restrict the meaning of the inverse function. Thus we restrict $\sin^{-1} y$ to mean that angle *lying between* $\pm \dfrac{\pi}{2}$ which has y as its sine.

In general, if we have to prove any result about a function, we are quite safe in proving it for a single-valued function only.

§ 5. Definition of a Limit.

Consider now the functional equation

$$y = \frac{x^2 - 1}{x - 1}.$$

For every value of x, except one, a single value of y can be found without difficulty, being in fact the same as the value of $x + 1$. The function $\dfrac{x^2 - 1}{x - 1}$ has thus a well-defined value for every value of x, except one.

The value excepted is evidently unity; for, if $x = 1$, numerator and denominator both vanish and we are left with $\dfrac{0}{0}$, which, of course, is meaningless. We cannot even get out of the difficulty by dividing out the common factor and leaving $y = x + 1$. For by the ordinary principles of Algebra, division by the zero factor $x - 1$ is invalid. In fact the expressions $\dfrac{x^2 - 1}{x - 1}$ and $x + 1$ are only identical just so long as we exclude the possibility $x = 1$.

We thus see that for the value $x = 1$ the function $\dfrac{x^2 - 1}{x - 1}$ assumes an "indeterminate" form, and the function cannot be said to have a value at all when $x = 1$, in the ordinary sense of the term "value."

Let us try another method of approach. Take a series of values of x converging down to unity, and evaluate y for each of these values. The results are tabulated as follows:

x	1·1	1·01	1·001	and so on
y	2·1	2·01	2·001	

Again, take a series of values below unity and converging up to it and we get the following table:

x	·9	·99	·999	and so on
y	1·9	1·99	1·999	

We see that the closer x approaches the value 1, the closer y approaches 2, whether we come down to the value $x = 1$ from above, or up to it from below. In fact, however close we may wish y to be to 2, we can find a value of x close to unity which will give us this value. For suppose we wish to have y within ·000001 of 2; it is necessary merely to take x between 2·000001 and 1·999999 and the required standard is attained.

Summing up, we see that the following facts are true:

(1) As x approaches unity, either from above or below, the values of y become closer and closer to 2.

(2) However close to 2 we may wish y to be, a value of x can be found near unity to give the desired result.

We express this state of affairs shortly by saying that

the "Limiting value" or "Limit" of y as x "approaches" or "tends to" 1 is 2. This is written in the form

$$\underset{x \to 1}{\mathrm{Lt}} \frac{x^2 - 1}{x - 1} = 2.$$

§ 6. Limits and Values.

The reader should make his mind quite clear on this subject before proceeding further. "Limits" should not be confused with values, and for this reason the term "Limit" is to be preferred to the alternative "Limiting value." The function never actually assumes the value 2, but can be made to approach it as closely as we like. The reader will probably have encountered a similar state of affairs in dealing with an infinite G.P. (though in this case n, the number of terms, is not a continuous variable).

It may perhaps be found easy to consider the question as if it were a game played by two opponents. First of all, your opponent states his standard of accuracy. He says: "I will accept the limit as 2 if you can get y within ·001 of 2." You counter him by saying "This can be done by taking x between ·999 and 1·001." He then changes his ground and says "I will increase the standard of accuracy to ·00001." You can then counter him by saying "Let x lie between ·99999 and 1·00001." And so on.

If you can always win, however stringent the standard of accuracy imposed, then the limit is 2, as you state it to be. But if at any stage your opponent defeats you, the limit is not 2.

This idea cannot easily be put into precise mathematical language without making the definition rather involved. But, for interest, we give here a complete definition of a limit.

Given a small number ε and two definite numbers a and A and that f(x) ∼ A < ε, if it is possible to find another small number η such that x ∼ a < η, however small ε may be; and conversely if for all values of x lying within η of a the difference f(x) ∼ A < ε, then A is said to be the limit of the function as x approaches a.

But the reader will probably find the idea easier to remember than the definition.

Note that it is necessary for $f(x)$ to approach the *same* number A whether x is just greater or just less than a; otherwise the function undergoes a sudden change in value and is said to be discontinuous. But we will revert to this subject a little later.

A similar mode of argument is seen to hold for a function which "increases without limit" or "tends to infinity" as x tends to some particular value. It is, of course, a contradiction in terms to speak of "infinity" as a "value" of the function, for this implies that "infinity" is a definite number.

Here again your opponent states his standard, which in this case is a very large number, let us say, 10000. You then have to prove the existence of a value of x close to the particular value mentioned such that $f(x) > 10000$, and so on. If this can be done, however large a number your opponent chooses, then the function is said to "become infinite" or "tend to infinity" at the value of x concerned. This state of affairs is written

$$\operatorname{Lt}_{x \to a} f(x) \to \infty$$

or simply $\qquad f(x) \to \infty$ as $x \to a$.

In many cases however the function changes sign as x passes through the value a, and this, and other reasons we shall consider in § 11, cause an "infinity" to be regarded as a "discontinuity" of a function.

§ 7. Rules for use of Limits.

We give here, without proof, the main rules for working with limits. It can be seen that the proof of these rules is somewhat beyond the scope of this work, but it can be safely assumed that these results are true for any functions which the student is likely to encounter in the course of ordinary work.

I. *The limit of the algebraic sum of any finite number of functions is equal to the like algebraic sum of their separate limits.*

II. *The limit of the product of any finite number of functions is equal to the product of their separate limits.*

III. *The limit of the quotient of two functions is equal to the quotient of their limits, provided that the limit of the divisor is not zero.*

A fourth rule, whilst not exactly fundamental, is usually assumed to hold.

IV. *The limit of the logarithm of any function is equal to the logarithm of the limit of the function.*

§ 8. An important Limit.

We give here as a worked example a very important Trigonometrical Limit, viz.

$$\underset{\theta \to 0}{\mathrm{Lt}} \frac{\sin \theta}{\theta}.$$

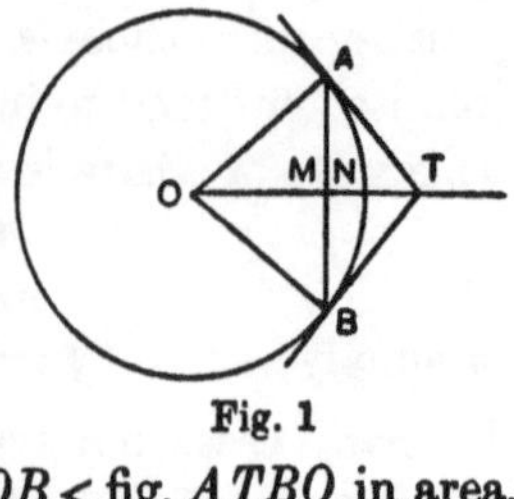

In Fig. 1 let the angle AOB be θ radians, ON bisecting the angle AOB.

Fig. 1

Then it is clear from elementary geometry that $\triangle AOB <$ sector $AOB <$ fig. $ATBO$ in area.

I.e. $\quad \frac{1}{2} OA^2 \sin \theta < \frac{1}{2} OA^2 . \theta < OA^2 . \tan \frac{\theta}{2},$

i.e. $\quad\quad\quad \sin \theta < \theta < 2 \tan \frac{\theta}{2}.$

Dividing both sides by $\sin \theta$,

$$1 < \frac{\theta}{\sin \theta} < \frac{2 \tan \frac{\theta}{2}}{\sin \theta},$$

i.e. $\quad\quad\quad 1 < \frac{\theta}{\sin \theta} < \sec^2 \frac{\theta}{2},$

or $\quad\quad\quad 1 > \frac{\sin \theta}{\theta} > \cos^2 \frac{\theta}{2}.$

Hence $\frac{\sin \theta}{\theta}$ always lies between 1 and $\cos^2 \frac{\theta}{2}$.

Now, by a proper choice of a small angle θ, $\cos^2 \frac{\theta}{2}$ may be made to approach unity as closely as we please. Hence the ratio $\frac{\sin \theta}{\theta}$ may be made to differ from unity by as little as we please by a proper choice of θ.

Hence, by definition,

$$\underset{\theta \to 0}{\text{Lt}} \ \frac{\sin \theta}{\theta} = 1.$$

This is an important result and should be carefully noted.

We have above

$$1 > \frac{\sin \theta}{\theta} > \cos^2 \frac{\theta}{2}.$$

If $\theta = 2\phi$, then $\quad 1 > \frac{\sin 2\phi}{2\phi} > \cos^2 \phi.$

Dividing by $\cos^2 \phi$, $\quad \sec^2 \phi > \frac{\tan \phi}{\phi} > 1.$

Hence, as above, $\quad \underset{\phi \to 0}{\text{Lt}} \ \frac{\tan \phi}{\phi} = 1.$

It should be noted that while $\dfrac{\sin \theta}{\theta}$ and $\dfrac{\tan \theta}{\theta}$ tend to the same Limit, one tends to it from below, the other from above.

§ 9. Sum of a Series: the Exponential Series.

The idea of the "sum" of an infinite series is similar to that of a Limit. If s_n, the sum of n terms of the series, tends to a limit S as $n \to \infty$, the series is said to "Converge" to this "sum." Tests for convergency of infinite series are found in text-books on Algebra. The reader is probably familiar with this notion, possibly through the "Geometric" series.

Many important convergent series are in general use and it is assumed that the reader has met the Binomial Series.

Another important convergent series which occurs is the Exponential Series. The infinite series

$$1 + x + \frac{x^2}{2!} + \frac{x^3}{3!} + \dots + \frac{x^n}{n!} + \dots$$

may be shown to be convergent for all values of x and is known as the "exponential" function of x. In particular the value assumed by this series when $x = 1$, viz.

$$1 + \frac{1}{1!} + \frac{1}{2!} + \dots + \frac{1}{n!} + \dots \text{ ad inf.,}$$

may be shown to lie between 2 and 3 and is defined as the number $e \ (= 2 \cdot 7183 \dots)$.

If $E(x)$, $E(y)$ denote the exponential functions of x and y, it may be shown that $E(x) \cdot E(y) = E(x + y)$ and thence that $E(x) = (E(1))^x = e^x$ for all values of x*.

* See text-books on Higher Algebra.

The series

$$1 + x + \frac{x^2}{2!} + \dots + \frac{x^n}{n!} + \dots \text{ ad inf.}$$

is therefore usually denoted by e^x.

It may also be shown that $e^x = \underset{n \to \infty}{\text{Lt}} \left(1 + \frac{x}{n}\right)^n$ and, in particular, that $e = \underset{n \to \infty}{\text{Lt}} \left(1 + \frac{1}{n}\right)^n$. This is sometimes taken as the definition of e.

§ 10. Continuity.

One further property of functions calls for notice before we pass on to the consideration of those processes which properly form the Differential Calculus. This property is that of Continuity.

We have noted above in dealing with the function $\frac{x^2 - 1}{x - 1}$ that it is necessary that the function should approach the same limit from either side of the value of x considered. If this is not the case the function is not "continuous." We may, in fact, take it as a broad general principle of Mathematics, or at any rate of functionality, that abrupt changes of any sort are unnatural, and gradual change the usual order of things. The easiest (though perhaps not quite the most accurate) definition of continuity is as follows:

A function of x is said to be continuous between two values a and b of x, if, starting from any value of x between a and b, a small change in the argument always produces a small change in the value of the function.

We might perhaps put the matter a little more rigidly thus: *If a small change in the argument produces everywhere a small change in the function, and if any change in the value of the function, however small, can be attributed to a small change in the argument, then the function is continuous over the range of values considered.*

§ 11. Discontinuities of a Function.

If at any particular value of x the condition of continuity is not satisfied, the function is said to have a "discontinuity" at that value. One type of discontinuity that may be possessed by a function is at once obvious; to wit, an "infinity." For in the neighbourhood of a value of x for which the function increases without limit ($\to \infty$) a small change in the value of the argument may make a very large change in the value of the function (cf. $\tan \theta$ in the neighbourhood of $\theta = \dfrac{\pi}{2}$. If θ is just less than $\dfrac{\pi}{2}$, $\tan \theta$ is very large and positive, but if θ is just greater than $\dfrac{\pi}{2}$ $\tan \theta$ is very large and negative).

Another type of discontinuity which a function may have is shown in Fig. 2. In this case the function approaches different limits from above and below the value of x represented by ON. This value of x is thus a discontinuity of the function. This type of discontinuity is rare, but it would be exhibited by the velocity-time diagram of a body which at some instant received a sudden impulse or restraint.

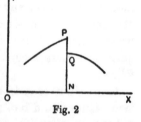

Fig. 2

§ 12. A Property of Continuous Functions.

The following is an important property of a continuous function.

If a function of x is continuous for every value of x lying between $x = a$ and $x = b$, and if $f(a) = A$ and $f(b) = B$; then, as x passes from a to b, $f(x)$ will take up every value between A and B at least once.

This is a difficult property to prove by a purely analytical argument. If, however, we confine our attention to functions that can be represented graphically (and, as we have pointed out, all the functions usually encountered can be so represented), then the property is evident from a figure. The reader should construct a figure for himself, and notice also that the converse of the theorem is not necessarily true.

Although no proof is given, the reader may safely assume that all the functions he is likely to encounter in the course of ordinary work will be continuous, except for a few special values of x. Infinities are the only kind of discontinuity likely to occur, and these are usually easily detected.

Further than this, any functions derived from continuous functions by the ordinary methods of Algebra, and also by the process of Differentiation given in later chapters, will themselves, in the main, be continuous. These latter functions may, however, have a limited number of discontinuities not possessed by the functions from which they are derived.

§ 13. Worked Examples on Limits.

We conclude the chapter with a few worked examples. In general the evaluation of limits requires a knowledge of the expansion of the functions concerned as series in powers of x; a matter with which we shall deal in Chap. VI. Lines of attack may, however, be suggested by the following three examples:

(1) Find $\displaystyle \operatorname*{Lt}_{n \to \infty} \frac{1^2 + 2^2 + \ldots + n^2}{n^3}$.

We know from Algebra that the numerator is

$$\frac{n(n+1)(2n+1)}{6}.$$

We therefore require

$$\operatorname*{Lt}_{n \to \infty} \frac{n(n+1)(2n+1)}{6n^3},$$

i.e.

$$\operatorname*{Lt}_{n \to \infty} \frac{1 \left(1 + \dfrac{1}{n}\right)\left(2 + \dfrac{1}{n}\right)}{6},$$

$$= \frac{1}{3}.$$

Notice that this example does not really contradict rule I of § 7, although the final limit is not the sum of the separate limits $\dfrac{1}{n^3}$, $\dfrac{2}{n^3}$, etc.

For the rule specifies a *finite* number of functions, and here the number is eventually infinite.

(2) $\displaystyle \operatorname*{Lt}_{x \to 1} \frac{x^{\frac{3}{2}} - 1}{x^{\frac{5}{2}} - 1}$.

Let $x = 1 + h$, then $x \to 1$ as $h \to 0$. Hence we require

$$\operatorname*{Lt}_{h \to 0} \frac{(1+h)^{\frac{3}{2}} - 1}{(1+h)^{\frac{5}{2}} - 1},$$

i.e.

$$\operatorname*{Lt}_{h \to 0} \frac{1 + \frac{3}{2}h + \frac{3}{8}h^2 + \ldots - 1}{1 + \frac{5}{2}h + \frac{15}{8}h^2 + \ldots - 1},$$

i.e.

$$\operatorname*{Lt}_{h \to 0} \frac{\frac{3}{2} + \frac{3}{8}h + \ldots}{\frac{5}{2} + \frac{15}{8}h + \ldots},$$

$$= \frac{3}{5}.$$

(3) Find $\qquad \underset{x \to \frac{\pi}{2}}{\text{Lt}} \left(\frac{\pi}{2} - x\right) \tan x.$

Let $x = \frac{\pi}{2} - h.$ $\quad \therefore$ as $x \to \frac{\pi}{2}$, $h \to 0.$

We now have to find $\quad \underset{h \to 0}{\text{Lt}} \, h \cot h,$

i.e. $\qquad\qquad\qquad \underset{h \to 0}{\text{Lt}} \, h \, \dfrac{\cos h}{\sin h},$

i.e. $\qquad\qquad\qquad \underset{h \to 0}{\text{Lt}} \, \dfrac{\cos h}{\dfrac{\sin h}{h}},$

$\qquad\qquad\qquad\qquad = 1$, applying Th. III of § 7.

EXAMPLES I

Find the following :

(1) $\underset{n \to \infty}{\text{Lt}} \dfrac{1 + 2 + 3 + \ldots + n}{n^2}.$

(2) $\underset{x \to \infty}{\text{Lt}} \dfrac{2x^2 - 3x + 1}{3x^2 - 2x - 1}.$

(3) $\underset{x \to 1}{\text{Lt}} \dfrac{2x^2 - 3x + 1}{3x^2 - 2x - 1}$

(4) $\underset{x \to 0}{\text{Lt}} \dfrac{\sqrt{1 + x} - 1}{x}.$

(5) $\underset{x \to 0}{\text{Lt}} \dfrac{ax^2 + bx}{bx^2 + ax}.$

(6) $\underset{x \to \infty}{\text{Lt}} \dfrac{ax^2 + bx}{bx^2 + ax}.$

(7) $\underset{x \to 0}{\text{Lt}} \dfrac{\sin ax}{\sin bx}.$

(8) $\underset{\theta \to 0}{\text{Lt}} \dfrac{1 - \cos \theta}{\theta^2}.$

(9) $\underset{x \to 0}{\text{Lt}} \dfrac{e^x - e^{-x}}{x}.$

(10) $\underset{x \to 0}{\text{Lt}} \dfrac{e^x + e^{-x} - 2}{x^2}.$

(11) $\underset{x \to 0}{\text{Lt}} \dfrac{\tan^{-1} x}{x}.$

(12) $\underset{x \to 1}{\text{Lt}} \dfrac{x^m - 1}{x^n - 1}.$

(13) $\underset{x \to +\infty}{\text{Lt}} \, xe^{-x}.$

(14) $\underset{x \to \pm\infty}{\text{Lt}} \dfrac{f(x)}{ax^n}$, where $f(x)$ denotes the polynomial

$$ax^n + bx^{n-1} + \ldots.$$

(15) Use No. 14 to show that for very large values of x either positive or negative, $f(x)$ may be replaced by $ax^n (1 + d)$ where d is a small number which $\to 0$ as $x \to \pm\infty.$

CHAPTER II

THE PROCESS OF DIFFERENTIATION

§ 14. Dynamical Considerations.

The best notion of the fundamental principles of the Differential Calculus can be obtained from everyday considerations of a dynamical nature. The conceptions thus developed can easily be extended to many branches of mathematics which are not dynamical but to which the same ideas are applicable.

We are all familiar with the expression "velocity" as applied to a moving body: in fact we are, for the most part, so familiar with it that we tend to overlook a difficulty in its meaning, which, if rightly understood, will give us the clue to the meaning of the fundamental process of the Differential Calculus.

What, then, do we mean if we state that the "velocity" of a moving object is 30 miles per hour, or 40 feet per second, and so on? We really mean that, if we took some interval of time, and measured the distance travelled by the moving body in that time, we should obtain by division a fraction $\frac{\text{Distance}}{\text{Time}}$. This fraction will represent the average distance traversed in a unit of time; and is defined as the "average velocity" of the moving object over the stated interval of time.

Now, in general, the interval over which we measure is a matter of our own choice. We might take a minute, or a second, or a quarter of an hour. But as long as the moving body is moving uniformly the fraction derived from these observations would be the same, provided we express each fraction in the same units. Thus, in the case of a body moving uniformly, there is no difficulty in defining its "velocity" at any point. It is simply its "average velocity" as found above.

§ 15. Non-uniform Motion.

But as soon as we begin to think of any change taking place in the motion, e.g. slowing down, speeding up, etc., we shall get widely differing results for our "average velocity." For instance, in slowing down we might observe the following table

Distance in feet	40	70	90	100	110
Time in seconds	1	2	3	4	5

giving five different "average velocities," viz. 40, 35, 30, 25, 22 feet per second, and it is now clear that, in the general case, the choice of the interval of time makes a great deal of difference. We can see clearly the safest plan to adopt in order to get the "velocity" *at* a given instant. We must make the interval of time as small as we conveniently can, hoping that the changes introduced in this small period may be so small as to make no serious error in the result. It is, in fact, clear that, errors in observation excluded, the smaller the interval of time we take, the more accurate will be our estimate of the result.

Putting the process into mathematical language, what we actually do is *to find the average velocity over a small interval of time and to find the limiting value to which this tends as the interval of time is made smaller and smaller.*

§ 16. Algebraical statement of the foregoing.

If we wish to express details concerning motion, we have to do it by means of an equation connecting s (the distance described) and t (the time taken) both expressed in suitable units and measured from some convenient starting-point. We can then assume that the motion is described by an equation

$$s = f(t).$$

To fix our ideas, let us consider the special case of the motion described by the equation

$$s(t) = t^3 + 3t^2 + 6t + 4,$$

where the notation $s(t)$ is to remind us that s, the distance, is dependent solely on t. We desire to find the velocity at some fixed time, say, a seconds after the commencement.

Referring to the previous article we choose a small interval of time, which we will denote by h seconds. We can now find the distance described in this interval by subtracting the distance described up to the beginning of this interval from that described up to the end of it.

Thus $\quad s(a+h) = (a+h)^3 + 3(a+h)^2 + 6(a+h) + 4,$

$\qquad\quad s(a) \quad = a^3 \qquad + 3a^2 \qquad + 6a \qquad + 4.$

$\therefore \quad s(a+h) - s(a) = 3a^2h + 3ah^2 + h^3 + 6ah + 3h^2 + 6h.$

$\therefore \quad$ the average velocity $=$ distance described $\div$ time taken

$$= \frac{s(a+h) - s(a)}{h}$$

$$= 3a^2 + 3ah + h^2 + 6a + h + 6$$

$$= (3a^2 + 6a + 6) + h(3a+1) + h^2.$$

The average velocity is thus clearly dependent on the value of the interval h. We now have to consider what happens when the interval h is diminished.

We should first notice that we cannot at once put $h = 0$, for we have previously performed an algebraical division by h, and division by a zero factor is invalid.

But as we have seen in the first chapter we have to get over this difficulty by using the Theory of Limits.

The average velocity $(3a^2 + 6a + 6) + h(3a + 1) + h^2$ clearly approaches $3a^2 + 6a + 6$ both from above and below as $h \rightarrow 0$, and also we can clearly find a value of h such that the difference $h(3a + 1) + h^2$ may be as small as we please.

Hence the limiting value of the average velocity is $3a^2 + 6a + 6$ and we take this to be the actual velocity at the instant when $t = a$.

It is also clear that $h(3a^2 + 6a + 6)$ is *approximately* equal to the distance described in the interval of time h seconds.

§ 17. Differentiation. Rate of Increase.

The performance of this operation is clearly not restricted to mechanical examples. It is, equally, the method by which we can find the "instantaneous" (as opposed to "average") rate of increase of any varying expression in terms of its variable. The usual name given to the whole process is "Differentiation"; and to the result of the process various names are given, the most common being "differential coefficient," "derivative," "derived function," "gradient" and so on. The reasons for these various names will be seen later.

We therefore see that the rule for finding the "differential coefficient" of a function $y(x)$, with respect to its variable x, can be expressed symbolically in the form

$$\underset{h \rightarrow 0}{\text{Lt}} \ \frac{y(x+h) - y(x)}{h}.$$

Very often we employ instead of h the slightly confusing notation δx (read as "delta-x") to denote a small increment in the value of x. We can then use δy for the corresponding increment in the value of y and put the rule in a shorter form thus:

$$\underset{\delta x \,\to\, 0}{\text{Lt}} \frac{\delta y}{\delta x}.$$

[It should be noticed that δ has no meaning by itself, but that δx together signify the choosing of a small increase in the value of x; and so for y.]

This mode of expression has served to introduce the standard notation for a differential coefficient, viz. $\dfrac{dy}{dx}$.

This is read as "dee-y by dee-x," and it should be clearly understood that it is not an algebraical fraction—for, since δy and δx become zero in the limit, the fraction $\dfrac{\delta y}{\delta x}$ has no meaning.

§ 18. Other Types of Notation.

To avoid this confusion of ideas, many writers have adopted the notation $\left(\dfrac{d}{dx}\right) y$; where $\dfrac{d}{dx}$ is regarded as a symbol of an operation to be performed on y, in the same way that $+$, $\times$, $\div$ are symbols of operation, and have no meaning if they stand alone.

Other commonly used notations are $D_x y$, y_x, $\dot{y}$, y_1, y', etc. The choice of notation is largely immaterial, as long as the reader remembers that, whatever notation we use, it represents a process performed on y and not a fraction.

To save the labour of performing the process *ab initio* for each example, a series of rules and standard forms has grown up, to which the name "Differential Calculus" has been given.

§ 19. Graphical Interpretation.

It is only reasonable to expect that, if the function $y(x)$ is capable of being represented graphically, as are almost all the functions we meet, the process which we have described in the preceding articles will have a definite meaning in relation to the figure.

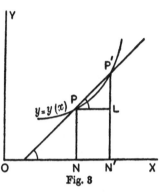

Fig. 3

In Fig. 3 let P be a point of the curve at which $x = a$, and let P' be the point at which $x = a + h$.

Now
$$NN' = PL = h.$$

Also
$$y(a) = PN,$$
$$y(a + h) = P'N'.$$
$$\therefore \quad y(a + h) - y(a) = P'L,$$

i.e.
$$\frac{y(a + h) - y(a)}{h} = \frac{P'L}{PL} = \tan P'PL,$$

i.e. is the tangent of the inclination of the chord PP' to the positive direction of the axis of x.

Now in the limit as $h \to 0$, the chord PP' tends to become the tangent to the curve at P. Hence

$$\underset{h \to 0}{\text{Lt}} \frac{y(a + h) - y(a)}{h}$$

finally becomes equal to the tangent of the angle of inclination of the tangent to the curve at P.

This angle is usually denoted by the Greek letter ψ (Psi) and the result can be written in the form

$$\frac{dy}{dx} = \tan \psi.$$

The expression "gradient of the curve at the point where $x = a$" is also used.

If y is a many-valued function of x, i.e. if there are several points on the curve for which $x = a$, there will be an appropriate value of $\frac{dy}{dx}$ for each of these points and in general these values will be different.

It should be noted that it is assumed throughout that the units are the same on both x- and y-axes. If this is not the case the tangent of the actual angle measured in the figure is not the value of $\frac{dy}{dx}$ at the point; but the reader should have no difficulty in seeing how the actual value may be arrived at.

If the curve falls between P and P', the corresponding value of $\tan \psi$ is negative and thus $\frac{dy}{dx}$ will be negative. And, in general, the ideas we have developed apply both to increasing and decreasing functions; since the only fundamental difference between increase and decrease is one of sign.

§ 20. Angle of Intersection of Two Curves.

We can at once find the angle at which two curves intersect. For if we find the coordinates of their points of intersection we can calculate $\frac{dy}{dx}$ for both curves and thus find $\tan \psi_1$ and $\tan \psi_2$.

If now α is the angle between the curves

$$\tan \alpha = \frac{\tan \psi_1 \sim \tan \psi_2}{1 + \tan \psi_1 \tan \psi_2},$$

and the solution can at once be completed by Trigonometrical methods.

§ 21. Acceleration.

Reverting again to the mechanical considerations with which we opened the chapter it becomes clear that "acceleration" stands in the same relation to velocity as velocity does to space described, i.e. acceleration is the rate of increase of velocity, and thus:

$$\text{Acceleration} = \frac{dv}{dt} = \frac{d}{dt}\left(\frac{ds}{dt}\right).$$

This latter expression is called the "second differential coefficient of s with respect to t," and is written

$$\frac{d^2s}{dt^2} \text{ or } \left(\frac{d}{dt}\right)^2 s.$$

Thus in § 16 we proved that the velocity at any time is given by the equation

$$v(t) = 3t^2 + 6t + 6,$$
$$\therefore \quad v(a) = 3a^2 + 6a + 6,$$
$$v(a+h) = 3(a+h)^2 + 6(a+h) + 6.$$

Hence the average acceleration

$$= \frac{v(a+h) - v(a)}{h} = 6a + 6 + h,$$

and the actual acceleration

$$= \underset{h \to 0}{\text{Lt}} \frac{v(a+h) - v(a)}{h} = 6a + 6,$$

i.e. the acceleration at time $t = 6t + 6.$

§ 22. Constant Terms.

The reader may have noticed that in each of these examples the constant term has given rise to nothing in the result; and we thus arrive at the rule: *The differential coefficient of a constant standing alone is zero.*

This stands evidently to reason—for, by definition, a constant is a number or expression which is not affected by a change in the variable and hence has no rate of increase for the variable.

§ 23. Worked Examples.

(1) *Find the equations to the tangents to the curve $y = x^2 + x + 1$ at the points where $y = 7$.*

Putting $y = 7$ in the equation, we get

$$x^2 + x - 6 = 0; \text{ i.e. } x = 2 \text{ or } -3;$$

$$\frac{dy}{dx} = \operatorname*{Lt}_{h \to 0} \frac{1}{h} \{(x+h)^2 + (x+h) + 1 - x^2 - x - 1\}$$

$$= \operatorname*{Lt}_{h \to 0} \frac{1}{h} \{2xh + h^2 + h\}$$

$$= 2x + 1.$$

Hence when $\qquad x = 2, \quad \dfrac{dy}{dx} = 5,$

and when $\qquad x = -3, \quad \dfrac{dy}{dx} = -5.$

In the first case we write down the equation of a line through $(2, 7)$ at an inclination 5.

Hence the first tangent is

$$y - 7 = 5(x - 2),$$
$$\text{i.e.} \qquad y = 5x - 3.$$

Similarly the second tangent is

$$y - 7 = -5(x - 2),$$
$$\text{i.e.} \qquad y = -5x + 17.$$

(2) *A right circular cylinder has a constant height c, but the radius of its base varies. If V is the volume and S the curved surface of the cylinder prove that $\dfrac{dV}{dr} = S$ and interpret the result geometrically.*

By formula $V = \pi r^2 c,$

$$\therefore \quad \frac{dV}{dr} = \operatorname*{Lt}_{h \to 0} \frac{1}{h} \{\pi c (r + h)^2 - \pi c r^2\}$$

$$= \operatorname*{Lt}_{h \to 0} \frac{1}{h} \{\pi c \cdot \overline{2rh + h^2}\}$$

$$= 2\pi c r$$

$$= \text{the curved surface.}$$

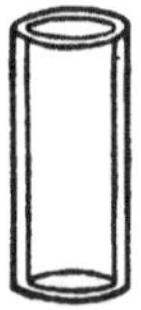

Fig. 4

Geometrically it is evident that a small increase in the radius increases the volume by a thin hollow cylinder as in diagram. When the increase in the radius is very small the volume of this hollow cylinder is approximately equal to its thickness multiplied by the area of either of the enclosing surfaces, one of which (the interior) is the cylinder;

i.e.   approximately $\delta V = h \times \text{surface}^*$,

i.e.   „ $\dfrac{\delta V}{h} = \text{surface}.$

Hence in the limit $\dfrac{dV}{dr} = \text{surface of cylinder.}$

(3) *To examine the variations in the value of the function*

$$x^3 - 3x^2 + 4.$$

Put $y = x^3 - 3x^2 + 4.$

* In general, if $\dfrac{\delta y}{\delta x} \to \dfrac{dy}{dx}$ as $x \to 0$, it also follows that $\delta y = \dfrac{dy}{dx} \cdot \delta x$ approximately. This transformation is often required in examples of small changes.

By consideration of previous results it is seen that

$$\frac{dy}{dx} = 3x^2 - 6x = 3x\,(x-2).$$

Now by definition $\frac{dy}{dx}$ measures the rate of increase (or decrease) of y in regard to x.

Clearly $\frac{dy}{dx}$ is positive when $x > 2$.

Also $\frac{dy}{dx}$ is positive when $x < 0$,

and $\frac{dy}{dx}$ is negative when $2 > x > 0$.

Therefore for all values except those between 0 and 2 y is an increasing function, and for those values it is a decreasing function.

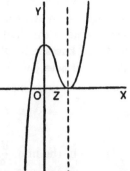

Fig. 5

Also by substitution when

$$x \to -\infty,\ y \to -\infty,$$
$$x = 0,\ y = 4,$$
$$x = 2,\ y = 0,$$
$$x \to +\infty,\ y \to +\infty.$$

And the graph is seen to be as in the diagram (Fig. 5).

Since $\frac{dy}{dx} = 0$, when $x = 0$ or 2 the tangent to the graph at these points is horizontal and these are known as "turning points" (see Ch. IV).

Much information about the roots of an equation can often be gained in this way. For example, it is clear from the figure that the equation $x^3 - 3x^2 + 4 = 0$ has one negative root, and two coincident positive roots. But the reader will be able to grasp this idea more easily after reading the theory of Maxima and Minima in Ch. IV and the sections relating to Rolle's Theorem in Ch. VI.

EXAMPLES II

Find from first principles the differential coefficients of the following :

(1) x^2, x^5, x^6.   (2) $\dfrac{1}{x}$, $\dfrac{1}{x^2}$, $\dfrac{1}{x^3}$.

(3) $\sqrt{x}$, $\sqrt[3]{x}$, $\sqrt{ax+b}$.   (4) $\dfrac{1}{\sqrt{x}}$, $\dfrac{1}{\sqrt{ax+b}}$.

Find the angles at which the following curves intersect :

(5) $y = x^2$, $y = x^5$.

(6) $y = 5 + x^2$, $y = 3(x+1)$.

(7) $y = x^2$, $y = \dfrac{1}{x}$.

(8) At what points of the curve $y = x^2 + 2x$ is the tangent

     (1) parallel to the x-axis,

     (2) equally inclined to the axes,

     (3) inclined at 30° to the (positive) x-axis?

(9) Write out the equations of the tangents in each case of the last example.

(10) At what points lying on the same ordinate are the two curves $y = x^3 - x$, $y = 4x - x^2$ parallel?

(11) Find the equation of the tangent to $y = x^3 + x + 1$ which is parallel to the line $3y + x = 0$.

(12) A particle moves in a straight line in such a way that

$$s = t^3 - 3t^2 + 3t + 8.$$

Find expressions for its velocity and acceleration at any instant, and show that there is only one point at which it is instantaneously at rest.

(13) Describe the character of the motion in Question 12.

(14) Prove that any motion (in a straight line) represented by an equation $s = At^2 + Bt + C$ where A, B, C are any constants is performed with constant acceleration. Find the acceleration and the initial velocity.

(15) If you are given an accurate graph showing the relation between s and t, explain how you would determine the velocity at any instant.

How could you decide from the figure whether acceleration or retardation was taking place at any given point?

(16) By consideration of its gradient, show that $x^2 + 5x + 11$ is always positive.

(17) By consideration of $\frac{dy}{dx}$ where $y = x^3 - 3x + 1$, prove that the equation $x^3 - 3x + 1 = 0$ has three real roots.

(18) Prove that $1 + x + x^2$ is an increasing function of x for all values of $x > -\frac{1}{2}$.

(19) For what range of values of x is $x^4 - 4x^3 + 4x^2 + 40$ (a) an increasing, (b) a decreasing, function of x?

(20) If A is the area of a circle of radius r, and C its circumference, prove that $\frac{dA}{dr} = C$.

Interpret this result geometrically.

(21) State and interpret a result similar to that of Question 20 for the sphere.

(22) The radius of a sphere is 4 inches, and at the moment of observation it is decreasing at $\frac{1}{8}$ inch per second.

Find the rate of decrease of (a) its volume, (b) its surface, at this instant.

N.B. $\frac{d}{dt} r^3$ may be taken to be $3r^2 \frac{dr}{dt}$.

(23) A hollow cone of height 20 inches and base radius 15 inches is filled with water and inverted. The water runs out through a hole at the vertex at a rate of 10 cubic inches per second. Find the rate at which the surface-area inside the cone is decreasing when the depth has dropped to 12 inches.

(24) A 6-foot man walks directly away from a 12 foot lamp-post at a rate of 6 feet per second. At what rate does his shadow lengthen?

(25) The area of a circle increases uniformly at a rate of 25 sq. cms. per second. At what rate is the radius increasing when the circle has reached a radius of 1 metre?

Does the radius also increase uniformly?

(26) The hypotenuse and one side of a right-angled triangle are measured and found to be 15 inches and 8 inches respectively. Each measurement is liable to an error of 1 °/₀ in either direction. Prove that the calculated length of the other side is liable to an error of as much as very nearly 1·8 °/₀.

*(27) Two particles are describing the curves $y = 4x^2 - 5x + 1$, $2y = x^2 - 3x + 4$ in such a way that the line joining them is always parallel to the y-axis. At what points will they be moving parallel to one another? At what rate will the distance between them be increasing at this instant?

(28) Find the condition that $x^3 + 3ax^2 + 3bx + c$ shall be an increasing function for all values of x.

(29) Show that the tangent to the curve $y = 3x^2 - 5x^3$, at the point where $x = \frac{3}{10}$, passes through the origin.

(30) Sand is being poured out at the rate of 462 cu. ins. per second, and forms a heap in the shape of a right circular cone, whose vertical angle is 90°.

At what rate, in inches per second, is the height increasing after 21 seconds? (Take $\pi = \frac{22}{7}$.)

CHAPTER III

GENERAL RULES AND STANDARD FORMS

§ 24. Standard Forms.

It would clearly be inconvenient if, in every example we dealt with, we had to go through the process of the preceding chapter at length. To meet this contingency we shall devote this chapter to proving certain rules and formulae, which may hereafter be used in all examples.

§ 25. Differentiation of x^n.

By definition
$$\frac{dy}{dx} = \underset{h \to 0}{\text{Lt}} \frac{(x+h)^n - x^n}{h}$$

$$= \underset{h \to 0}{\text{Lt}} \, x^n \frac{\left(1 + \dfrac{h}{x}\right)^n - 1}{h}$$

$$= x^n \underset{h \to 0}{\text{Lt}} \frac{\left(1 + \dfrac{h}{x}\right)^n - 1}{h},$$

taking x^n outside the limit sign, which is valid since x^n does not contain h.

But since $\dfrac{h}{x}$ is by supposition small, we may use the Binomial Theorem, whence

$$\left(1 + \frac{h}{x}\right)^n = 1 + n\frac{h}{x} + \frac{n(n-1)}{2!}\frac{h^2}{x^2} + \dots,$$

the expansion being valid for all values of n.

$$\therefore \quad \frac{\left(1 + \dfrac{h}{x}\right)^n - 1}{h} = \frac{n}{x} + \frac{n(n-1)}{2!}\frac{h}{x^2} + \text{higher powers of } h.$$

$$\therefore \quad \underset{h \to 0}{\text{Lt}} \frac{\left(1 + \dfrac{h}{x}\right)^n - 1}{h} = \frac{n}{x}.$$

$$\therefore \quad \frac{dy}{dx} = x^n \frac{n}{x} = nx^{n-1}.$$

Hence for all values of **n**, the differential coefficient of x^n is nx^{n-1}.

§ 26. Differentiation of sin x.

Notice first that we mean x radians.

Here we have to find

$$\underset{h \to 0}{\text{Lt}} \frac{\sin(x+h) - \sin x}{h}$$

$$= \underset{h \to 0}{\text{Lt}} \frac{2 \cos\left(x + \dfrac{h}{2}\right) \sin \dfrac{h}{2}}{h}.$$

But by supposition h is a small angle, and thus

$$\underset{h \to 0}{\text{Lt}} \frac{2 \sin \dfrac{h}{2}}{h} = \underset{h \to 0}{\text{Lt}} \frac{\sin \dfrac{h}{2}}{\dfrac{h}{2}} \text{ which is } 1 \quad (\text{see § 8}).$$

Hence limit required is the limit of

$$\cos\left(x + \frac{h}{2}\right),$$

which is clearly $\cos x$.

The differential coefficient of **sin x** is **cos x** and the reader will find no difficulty in proving that **the differential coefficient of cos x is − sin x.**

§ 27. Differentiation of e^x.

Here we require $$\underset{h \to 0}{\text{Lt}} \frac{e^{x+h} - e^x}{h}$$

$$= e^x \underset{h \to 0}{\text{Lt}} \frac{e^h - 1}{h}.$$

But $$e^h - 1 \equiv h + \frac{h^2}{2!} + \frac{h^3}{3!} + \dots. \quad \text{(see § 9)}$$

$$\therefore \underset{h \to 0}{\text{Lt}} \frac{e^h - 1}{h} \text{ is } 1.$$

Hence **the differential coefficient of e^x is e^x itself.**

It can be proved that the only function whose derivative is equal to itself is a multiple of e^x. That is to say, the exponential function must enter into the composition of any function whose rate of increase at any point is equal (or proportional) to the function itself. Many such examples occur physically—and it is the occurrence of these examples which gives the study of the Exponential Function the importance it possesses in Higher Mathematics.

Before carrying our list of standard forms any further, we must first prove five general rules which are valid whatever the nature of the functions involved.

§ 28. General Rules for Differentiation.

A. *The differential coefficient of a constant standing by itself is zero.* This has already been noted in § 22.

B. *The differential coefficient of a function multiplied by a constant is the differential of the function, itself multiplied by a constant,* i.e.

$$\frac{d}{dx} C f(x) = C \frac{d}{dx} f(x).$$

This again is evident from the definition of the differential coefficient as a rate of increase.

C. *The differential coefficient of the sum (or difference) of a number of functions is the sum (or difference) of the separate differential coefficients*, e.g.

$$\frac{d}{dx}\{x^3 - \sin x + e^x - \cos x\}$$

$$= \frac{d}{dx} x^3 - \frac{d}{dx} \sin x + \frac{d}{dx} e^x - \frac{d}{dx} \cos x$$

$$= 3x^2 - \cos x + e^x + \sin x.$$

This result is evident from the idea of a rate of increase; it is also a special case of § 7, Rule I.

D. *The rule for differentiating a product is*

$$\frac{d}{dx}(UV) = U\frac{dV}{dx} + V\frac{dU}{dx}.$$

To prove this let $y = UV$ and suppose that when x increases to $x + h$, U increases to $U + u$ and V to $V + v$. Then u and v will both be small quantities involving h*.

The product will now be $(U + u)(V + v)$.

$$\therefore \quad \frac{dy}{dx} = \underset{h \to 0}{Lt} \frac{(U + u)(V + v) - UV}{h}$$

$$= \underset{h \to 0}{Lt} \frac{Uv + Vu + uv}{h}$$

$$= \underset{h \to 0}{Lt} U\frac{v}{h} + \underset{h \to 0}{Lt} V\frac{u}{h} + \underset{h \to 0}{Lt} \frac{uv}{h}.$$

Now since u and v are both small and comparable with h, the ratios $\frac{u}{h}, \frac{v}{h}$ will tend to finite limits; these will be in

* Observe that this necessitates the continuity of U and V.

fact, $\dfrac{dU}{dx}$ and $\dfrac{dV}{dx}$. But $\dfrac{uv}{h}$ will be comparable with $\dfrac{h^2}{h}$, i.e. it will tend to zero.

Hence we get, in the limit,

$$\frac{d}{dx}(UV) = U\frac{dV}{dx} + V\frac{dU}{dx}.$$

This rule can readily be applied to any number of functions; and the result expressed symbolically as follows:

$$\frac{d}{dx}(UVW\ldots) = VW\ldots\frac{dU}{dx} + UW\ldots\frac{dV}{dx}$$
$$+ UV\ldots\frac{dW}{dx} + \ldots.$$

This case can be dealt with more easily by a method given in § 35.

E. *Rule for a Quotient.*

In the same notation, the rule for differentiating a quotient is

$$\frac{d}{dx}\left(\frac{U}{V}\right) = \frac{V\dfrac{dU}{dx} - U\dfrac{dV}{dx}}{V^2}.$$

For here we have to evaluate

$$\operatorname*{Lt}_{h\to 0}\frac{1}{h}\left\{\frac{U+u}{V+v} - \frac{U}{V}\right\}$$
$$= \operatorname*{Lt}_{h\to 0}\frac{1}{h}\left\{\frac{(U+u)(V-v)}{V^2-v^2} - \frac{UV}{V^2}\right\}.$$

As before we can neglect v^2* and the expression becomes

$$\operatorname*{Lt}_{h\to 0}\frac{1}{h}\left\{\frac{UV + uV - vU - uv - UV}{V^2}\right\}$$
$$= \operatorname*{Lt}_{h\to 0}\frac{V\dfrac{u}{h} - U\dfrac{v}{h} - \dfrac{uv}{h}}{V^2},$$

* Since it involves at least the square of h.

which, neglecting $\dfrac{uv}{h}$ as above, reduces to

$$\frac{V\dfrac{du}{dx} - U\dfrac{dv}{dx}}{V^2}.$$

§ 29. Further Standard Forms.

We are now able to add to our standard forms the following useful results.

(1) Let $\qquad y = \tan x = \dfrac{\sin x}{\cos x},$

$$\frac{dy}{dx} = \frac{\cos x \cdot \dfrac{d}{dx}\sin x - \sin x \cdot \dfrac{d}{dx}\cos x}{\cos^2 x}$$

$$= \frac{\cos^2 x + \sin^2 x}{\cos^2 x}$$

$$= \frac{1}{\cos^2 x} = \sec^2 x.$$

Hence $\qquad \dfrac{d}{dx}\tan x = \sec^2 x,$

and it is easily proved in the same way that

$$\frac{d}{dx}\cot x = -\operatorname{cosec}^2 x.$$

(2) Let $\qquad y = \sec x = \dfrac{1}{\cos x},$

$$\frac{dy}{dx} = \frac{\cos x \cdot \dfrac{d}{dx}(1) - 1 \cdot \dfrac{d}{dx}\cos x}{\cos^2 x}$$

$$= \frac{\sin x}{\cos^2 x}.$$

Hence
$$\frac{d}{dx}\sec x = \frac{\sin x}{\cos^2 x},$$

and similarly
$$\frac{d}{dx}\operatorname{cosec} x = -\frac{\cos x}{\sin^2 x}.$$

The reader should make for himself a list of the more important standard forms as they occur, and, of course, commit them to memory. In many examples algebraic and trigonometric simplifications, such as division, etc., should be carried as far as possible before using the Calculus, as much needless labour is often avoided in this way. For example

$$\frac{x^2+1}{x+1} = x - 1 + \frac{2}{x+1},$$

$$\frac{x}{1+x} = 1 - \frac{1}{1+x},$$

$$\frac{2\tan\frac{x}{2}}{1-\tan^2\frac{x}{2}} = \sin x,$$

and so on.

EXAMPLES III A

(1) Prove the formulae for the differentiation of (a) $\cos x$, (b) $\cot x$, (c) $\operatorname{cosec} x$†.

Using the formulae given, differentiate the following :

(2) $3x^2+2x+6.$ (3) $\dfrac{1}{x^3} - \dfrac{2}{x^2} + \dfrac{3}{x}.$ (4) $\dfrac{x^2+x+1}{x}.$

(5) $\dfrac{x^3+1}{x+1}.$ (6) $(x^2-x+1)^2.$

(7) $3\sqrt{x} - 4\sqrt[3]{x} + 5\sqrt[4]{x}.$ (8) $(x+1)(\sqrt{x}-1).$

† An an aid to memory notice that the differential coefficients of all the co-functions have a negative sign.

(9) $\dfrac{x}{x+1}$.   (10) $\dfrac{3x}{4x+3}$.   (11) $\dfrac{x^2+4x+3}{x+4}$.

(12) $x \cos x$.   (13) $\dfrac{\cos x}{x}$.   *(14) $x^2 \sin x$.

(15) $\sin^2 x$ and deduce $\dfrac{d}{dx}(\cos^2 x)$.   (16) $\dfrac{\sin x}{x^3}$.

(17) $\tan^2 x$ and deduce $\dfrac{d}{dx}(\sec^2 x)$.   (18) $x \sin x \cos x$.

(19) $\sin^3 x$.

(20) Deduce from Exx. 15 and 19 the value of $\dfrac{d}{dx}\sin^n x$.

(21) $(x^2+1)^2$.   (22) $(x^2+1)\sec x$.   (23) $e^x \operatorname{cosec} x$.

(24) $e^x(\tan x + \sec x)$.   (25) $\sin^2\left(\dfrac{\pi}{4}+\dfrac{x}{2}\right)$.

*(26) $\dfrac{1+2\cos x}{2+\cos x}$.   *(27) $x - \dfrac{\sin x\,(14+\cos x)}{9+6\cos x}$.

(28) A triangle ABC has a fixed base AB and a constant vertical angle C. Prove that $\dfrac{da}{db} = -\dfrac{\cos A}{\cos B}$.

§ 30. Functions of a Function.

There is a sixth general rule for differentiation which is of frequent application. We often need to differentiate expressions, such as

$$\sqrt{x^2+a^2}, \qquad \sin 3x, \qquad \cos^3 x, \text{ etc.}$$

Such expressions are known as "double functions" or "functions of a function."

Considering a general case, suppose that Y is a function of U which is itself a function of x. (For example $\sin 3x$ is a function of $3x$ which is itself a function of x.)

As x increases by h, let U increase by u, and, as a result of this increase, let Y increase by y.

Then $\quad \dfrac{dY}{dx} = \underset{h \to 0}{\mathrm{Lt}}\ \dfrac{Y+y-Y}{h}$

$$= \underset{h \to 0}{\mathrm{Lt}}\ \dfrac{Y+y-Y}{U+u-U}\ \dfrac{U+u-U}{h}$$

$$= \underset{h \to 0}{\mathrm{Lt}}\ \dfrac{Y+y-Y}{U+u-U} \cdot \underset{h \to 0}{\mathrm{Lt}}\ \dfrac{U+u-U}{h}.$$

Now since u and h both tend to zero together we may re-write this last expression as

$$\underset{u \to 0}{\mathrm{Lt}}\ \dfrac{Y+y-Y}{U+u-U} \cdot \underset{h \to 0}{\mathrm{Lt}}\ \dfrac{U+u-U}{h},$$

i.e. as $\qquad\qquad \dfrac{dY}{dU} \cdot \dfrac{dU}{dx}.$

The result may readily be extended to functions of any degree of complexity: if, say, $y = \phi(u)$, $u = f(v)$, $v = \psi(x)$, then

$$\dfrac{dy}{dx} = \dfrac{dy}{du} \cdot \dfrac{du}{dv} \cdot \dfrac{dv}{dx},$$

and so on.

The student will find that, with a little practice, he is able to dispense with the auxiliary variable u and to write the result at once. Thus

$$\dfrac{d}{dx}(x^2 + 2x + 3)^{\frac{3}{2}} = \tfrac{3}{2}(x^2 + 2x + 3)^{\frac{1}{2}}(2x+2).$$

§ 31. Examples of Differentiation.

(1) $\qquad\qquad y = \sqrt{x^2 + a^2}.$

Here $\qquad u = x^2 + a^2 \quad$ and $\quad y = u^{\frac{1}{2}},$

$$\dfrac{du}{dx} = 2x, \qquad\qquad \dfrac{dy}{du} = \tfrac{1}{2}u^{-\frac{1}{2}}.$$

$$\therefore \quad \frac{dy}{dx} = \frac{dy}{du} \cdot \frac{du}{dx}$$

$$= \tfrac{1}{2} u^{-\frac{1}{2}} \cdot 2x$$

$$= \frac{x}{\sqrt{u}} = \frac{x}{\sqrt{x^2 + a^2}}.$$

(2) $\qquad\qquad y = \sin 3x.$

Here $\qquad\qquad u = 3x, \qquad y = \sin u,$

$$\frac{du}{dx} = 3, \qquad \frac{dy}{du} = \cos u.$$

$$\therefore \quad \frac{dy}{dx} = \frac{du}{dx} \cdot \frac{dy}{du} = 3 \cos u$$

$$= 3 \cos 3x.$$

(3) $\qquad\qquad y = e^{x \cos x}.$

Here $\qquad\qquad u = x \cos x, \qquad\qquad y = e^u,$

$$\frac{du}{dx} = \cos x - x \sin x, \quad \frac{dy}{du} = e^u.$$

$$\therefore \quad \frac{dy}{dx} = e^u (\cos x - x \sin x)$$

$$= (\cos x - x \sin x) \, e^{x \cos x}.$$

§ 32. Inverse Functions.

The method provides us with an easy means of differentiating inverse functions. For if we consider y as a function of x, and then x as a function of y,

$$\frac{dy}{dy} = \frac{dy}{dx} \cdot \frac{dx}{dy},$$

i.e. $\qquad\qquad 1 = \frac{dy}{dx} \cdot \frac{dx}{dy},$

or $\qquad\qquad \frac{dx}{dy} = \frac{1}{\dfrac{dy}{dx}}.$

We should notice that this result does not depend upon the rules of Elementary Algebra, for we have observed that dy and dx mean nothing by themselves, and hence cannot be "cancelled."

The result is, however, easily seen geometrically. For $\frac{dy}{dx}$ is the tangent of the angle between the tangent to the curve at some point and the axis of x, and by analogy $\frac{dx}{dy}$ is the tangent of its inclination to the axis of y. But, since these angles are complementary,

$$\frac{dy}{dx} \cdot \frac{dx}{dy} = \tan \psi \cdot \tan \left(\frac{\pi}{2} - \psi\right) = 1.$$

§ 33. More Standard Forms.

(1) Let
$$y = \log_e x.$$

Then
$$e^y = x.$$

$$\therefore \; \frac{dx}{dy} = e^y = x.$$

$$\therefore \; \frac{dy}{dx} = \frac{1}{x}.$$

Hence
$$\frac{d}{dx} \log_e x = \frac{1}{x}.$$

Also since
$$\log_a x = \log_e x \cdot \log_a e,$$

$$\frac{d}{dx} \log_a x = \frac{1}{x} \log_a e.$$

(2)
$$y = a^x,$$

i.e.
$$x = \log_a y.$$

$$\therefore \; \frac{dx}{dy} = \frac{1}{y} \log_a e.$$

$$\therefore \quad \frac{dy}{dx} = \frac{y}{\log_a e} = y \log_e a.$$

$$\therefore \quad \frac{d}{dx} a^x = a^x \log_e a.$$

(3) $$y = \sin^{-1} x.$$

$$\therefore \quad \sin y = x,$$

$$\frac{dx}{dy} = \cos y = \sqrt{1 - x^2}.$$

(The positive sign is attributed to the radical; for, as we have seen, in the equation $y = \sin^{-1} x$, y lies between $\pm \frac{\pi}{2}$, otherwise the function is not single-valued (§ 4). Hence $\cos y$ is positive.)

Hence $$\frac{d}{dx} \sin^{-1} x = \frac{1}{\sqrt{1 - x^2}},$$

and similarly $$\frac{d}{dx} \cos^{-1} x = - \frac{1}{\sqrt{1 - x^2}}.$$

(4) $$y = \tan^{-1} x.$$

$$\therefore \quad \tan y = x,$$

$$\frac{dx}{dy} = \sec^2 y = 1 + \tan^2 y = 1 + x^2.$$

$$\therefore \quad \frac{d}{dx} \tan^{-1} x = \frac{1}{1 + x^2},$$

and $$\frac{d}{dx} \cot^{-1} x = - \frac{1}{1 + x^2}.$$

In a similar way any other inverse functions may be differentiated.

§ 34. Standard Forms.

We give here for reference a short table of the standard forms already found, but the student is earnestly advised to make his own and to add to it new forms as found.

Form	Differential Coefficient
x^n	nx^{n-1}
$\sin x$	$\cos x$
$\cos x$	$-\sin x$
$\tan x$	$\sec^2 x$
$\cot x$	$-\operatorname{cosec}^2 x$
$\sec x$	$\dfrac{\sin x}{\cos^2 x}$ or $\sec x \tan x$
$\operatorname{cosec} x$	$-\dfrac{\cos x}{\sin^2 x}$ or $-\operatorname{cosec} x \cot x$
e^x	e^x
a^x	$a^x \log_e a$
$\log_e x$	$\dfrac{1}{x}$
$\sin^{-1} x$	$\dfrac{1}{\sqrt{1-x^2}}$
$\cos^{-1} x$	$-\dfrac{1}{\sqrt{1-x^2}}$
$\tan^{-1} x$	$\dfrac{1}{1+x^2}$
$\cot^{-1} x$	$-\dfrac{1}{1+x^2}$
$\sec^{-1} x$	$\dfrac{1}{x\sqrt{x^2-1}}$
$\operatorname{cosec}^{-1} x$	$-\dfrac{1}{x\sqrt{x^2-1}}$

EXAMPLES III B

Differentiate the following :

(1) $(1+2x)^3$.　　(2) $(1+2x^2)^3$.　　(3) $(a+bx^n)^2$.

(4) $(a+bx^n)^m$.　　(5) $\dfrac{1}{(1-3x)^2}$.　　(6) $\sin 3x - 2$.

(7) $\sin(3x-2)$.   **(8)** $\sin^3(x-2)$.   **(9)** e^{x^2}.

(10) e^{2x}.   **(11)** $\{e^{x^2}\}^3$.   **(12)** $\sqrt{a^2-x^3}$.

(13) $\sqrt{1+3x+5x^3}$.   **(14)** $\sqrt{a+bx+cx^3}$.   **(15)** $e^{\sqrt{a+bx+cx^3}}$.

(16) $(a+x)\sqrt{a^2+x^2}$.   **(17)** $\dfrac{x+1}{\sqrt{x^2+1}}$.

(18) $\sqrt[3]{x^3+\dfrac{3}{x^3}}$.   **(19)** $\sin^{-1}(e^x)$.   **(20)** $\log_e(\sin x)$.

(21) $\log_e\tan x$.   **(22)** $\sqrt{\dfrac{1-x}{1+x}}$.

(23) $\dfrac{\sqrt{1+x}-\sqrt{1-x}}{\sqrt{1+x}+\sqrt{1-x}}$.   *__(24)__ $\log_e(x+\sqrt{x^2+1})$.

(25) $\log_{10}(x^3+x+1)$.   **(26)** $e^{\sin^{-1}x}$.

(27) $e^{x\tan^{-1}x}$.   **(28)** $\sqrt{\log(x^2+1)}$.   **(29)** $\log\log x$.

(30) $\tan^{-1}\dfrac{x\sqrt{2}}{1-x^2}$.   **(31)** $\tan^{-1}\sqrt{\dfrac{1-e}{1+e}}\tan\dfrac{x}{2}$ $(e<1)$

*__(32)__ $\log\sqrt{\dfrac{x^2+\sqrt{2}x+1}{x^2-\sqrt{2}x+1}}$.   *__(33)__ $\log(\sec x+\tan x)$.

*__(34)__ $\tan^{-1}\sqrt{\dfrac{x-a}{x-b}}$.   **(35)** $\sin^{-1}\dfrac{x}{a}$.   **(36)** $\tan^{-1}\dfrac{x}{a}$.

(37) $\sec^{-1}x$.   **(38)** $\cos^{-1}(1-2x^2)$.

*__(39)__ $\sin^{-1}\dfrac{2x}{\sqrt{4+x^4}}$.   **(40)** $\cosh^{-1}x\left\{\cosh x=\dfrac{e^x+e^{-x}}{2}\right\}$.

(41) $\sinh^{-1}x\left\{\sinh x=\dfrac{e^x-e^{-x}}{2}\right\}$.

(42) Differentiate $\tan^{-1}x$ with respect to $\sin^{-1}x$.

(43) Differentiate ax^3+bx+c with respect to $\log_e x$.

(44) Differentiate $\cosh x$ with respect to e^x.

§ 35. Logarithmic Differentiation.

The form of the derivative of a logarithm gives rise to a method known as Logarithmic Differentiation, which is often of use in dealing with complicated products and powers, particularly where the index of the power contains the variable. It consists merely in taking logarithms before differentiating, e.g.

$$y = uvw \ldots,$$
$$\log y = \log u + \log v + \log w + \ldots.$$
$$\therefore \quad \frac{1}{y}\frac{dy}{dx} = \frac{1}{u}\frac{du}{dx} + \frac{1}{v}\frac{dv}{dx} + \frac{1}{w}\frac{dw}{dx} + \ldots *$$

treating each logarithm as a function of a function.

For example
$$y = \sin x \sin 2x \sin 3x.$$
$$\therefore \quad \log y = \log \sin x + \log \sin 2x + \log \sin 3x.$$
$$\therefore \quad \frac{1}{y}\frac{dy}{dx} = \frac{\cos x}{\sin x} + \frac{2 \cos 2x}{\sin 2x} + \frac{3 \cos 3x}{\sin 3x}.$$
$$\therefore \quad \frac{dy}{dx} = \sin x \sin 2x \sin 3x \{\cot x + 2 \cot 2x + 3 \cot 3x\}.$$

Or, again, consider $y = x^{\cos x}$.
$$\therefore \quad \log y = \cos x \log_e x.$$
$$\therefore \quad \frac{1}{y}\frac{dy}{dx} = -\sin x \log_e x + \cos x \cdot \frac{1}{x}.$$
$$\therefore \quad \frac{dy}{dx} = x^{\cos x} \left\{ \frac{\cos x}{x} - \sin x \log_e x \right\}.$$

§ 36. Trigonometrical Simplification.

In many cases, particularly of inverse trigonometrical functions, much labour can be saved by simplifying before differentiation. Many of these simplifications are rather difficult to foresee, and the student can only acquire them by practice. The following examples indicate some of the methods followed.

* Cf. end of § 28, D.

(1)
$$y = \tan^{-1} \frac{a - bx}{b + ax}$$

$$= \tan^{-1} \frac{\dfrac{a}{b} - x}{1 + \dfrac{a}{b} x}$$

$$= \tan^{-1} \frac{a}{b} - \tan^{-1} x.$$

$$\therefore \quad \frac{dy}{dx} = - \frac{1}{1 + x^2},$$

for $\dfrac{a}{b}$ is a constant.

(2)
$$y = \sin^{-1} \frac{2x}{1 + x^2}.$$

Let
$$x = \tan \frac{\theta}{2}.$$

Then
$$y = \sin^{-1} \frac{2 \tan \dfrac{\theta}{2}}{1 + \tan^2 \dfrac{\theta}{2}} = \sin^{-1} \sin \theta$$

$$= \theta$$

$$= 2 \tan^{-1} x.$$

$$\therefore \quad \frac{dy}{dx} = \frac{2}{1 + x^2}.$$

Algebraic division and rationalization should also be carried as far as possible before commencing to differentiate.

§ 37. Implicit Relations.

We often need to find $\dfrac{dy}{dx}$ from an equation which does not give y in terms of x directly, but which gives a relation between x and y. Such an equation is called an implicit equation (see § 3).

To fix our ideas, suppose we have to find $\dfrac{dy}{dx}$ from the equation $\qquad x^2 + y^2 + 2x + 4y - 6 = 0.$

All we have to do is to differentiate every term of the equation, remembering that y is a function of x and using the rule for a function of a function.

Thus $\qquad 2x + 2y\dfrac{dy}{dx} + 2 + 4\dfrac{dy}{dx} = 0.$

$$\therefore \quad \frac{dy}{dx} = -\frac{2x+2}{2y+4}.$$

It is, of course, not often possible to get a simple result for $\dfrac{dy}{dx}$ in terms of x alone in these cases.

§ 38. Partial Differential Coefficients.

We now notice that the two components of the fraction given above are exactly what we should have obtained if we had differentiated the equation first with respect to x, treating y as a constant, and then with respect to y, treating x as a constant.

This kind of differentiation is known as "partial" differentiation with respect to x or y and the two results are written as $\dfrac{\partial u}{\partial x}$ and $\dfrac{\partial u}{\partial y}$ respectively, where $u \equiv 0$ represents the equation.

We can thus state the rule for an implicit equation $u \equiv 0$ in the following form, which will be proved later,

$$\frac{dy}{dx} = -\frac{\dfrac{\partial u}{\partial x}}{\dfrac{\partial u}{\partial y}},$$

e.g. if $u \equiv ax^2 + 2hxy + by^2 - 1 = 0$,

$$\text{then } \frac{\partial u}{\partial x} = 2ax + 2hy,$$

$$\text{and } \frac{\partial u}{\partial y} = 2hx + 2by.$$

$$\therefore \quad \frac{dy}{dx} = -\frac{ax + hy}{hx + by}.$$

Partial Differentiation will be discussed at length in Chapter XII, where a proof of this rule is given.

EXAMPLES III C

Find $\frac{dy}{dx}$ in the following cases:

(1) $y = 2^{x^2}$. (2) $y = x^x$. (3) $y = x^{e^x}$.

(4) $y = (x^e)^x$. (5) $y = x^x + x^{2/x}$.

*(6) $y = \dfrac{\sqrt{x^2 + a^2} - \sqrt{x^2 - a^2}}{\sqrt{x^2 + a^2} + \sqrt{x^2 - a^2}}$.

(7) $y = \log \dfrac{x^2 - 2x + 1}{x^2 + 4x + 3}$.

(8) $y = \sin^{-1}\{a\sqrt{1 - x^2} - x\sqrt{1 - a^2}\}$. (9) $y = \cos^{-1}(2x^2 - 1)$.

(10) $y = \sin^{-1}(3x - 4x^3)$. (11) $y = \tan^{-1}\dfrac{3x - x^3}{1 - 3x^2}$.

*(12) $y = \dfrac{x + \sqrt{x^2 - 1}}{x - \sqrt{x^2 - 1}}$.

*(13) $y = \tan^{-1}\dfrac{\cos x - \sin x}{\cos x + \sin x}$.

Find $\frac{dy}{dx}$ in terms of x and y in the following cases:

(14) $(x + 3)^2 + (y + 4)^2 - 16 = 0$. (15) $b^2x^2 + a^2y^2 = a^2b^2$.

(16) $ax^2 + 2hxy + by^2 + 2gx + 2fy + c = 0$. (17) $x^{\frac{2}{3}} + y^{\frac{2}{3}} = a^{\frac{2}{3}}$.

(18) $xy^y = $ const.   (19) $x \sin y + y \sin x = 0$.

(20) $y = x^y$.   (21) $x^m y^n = (x+y)^{m+n}$.

*(22) $(x+y)^m x^n = (x-y)^n y^m$.

Calculate $x \dfrac{\partial u}{\partial x} + y \dfrac{\partial u}{\partial y}$ for the following functions:

(23) $u \equiv ax^2 + 2hxy + by^2$.   (24) $u \equiv (x^{\frac{1}{3}} + y^{\frac{1}{3}})(x+y)$.

(25) $u \equiv x^4 + x^3 y + 4x^2 y^2 + xy^3 + y^4$.

(26) $u \equiv \sin^{-1} \dfrac{\sqrt{x^3 + y^3}}{x+y}$.

(27) Comparing the results of the last four examples, write down an experimental formula for $x \dfrac{\partial u}{\partial x} + y \dfrac{\partial u}{\partial y}$. Test this on any functions of x and y and state when it is true and when it is not.

(28) If $x = \cos \theta$, $y = \sin^3 \theta$, express $\dfrac{d^2 y}{dx^2}$ in terms of θ.

(29) Find $\dfrac{d^2 y}{dx^2}$ if $x = at^2$, $y = at^3$.

*(30) Prove that, if $x = \cos \theta$, $y = \sin^7 \theta$, then $\dfrac{d^2 y}{dx^2} + \dfrac{105}{4} \sin 4\theta = 0$.

(31) Given $x = \dfrac{am}{1+m^3}$, $y = \dfrac{am^2}{1+m^3}$, show that $\dfrac{dy}{dx} = \dfrac{2m - m^4}{1 - 2m^3}$.

(32) Show that $y = A \cos 3x + B \sin 3x$ satisfies the equation

$$\frac{d^2 y}{dx^2} + 9y = 0.$$

(33) Prove that $y = e^x \cos x$ satisfies the equation

$$\frac{d^2 y}{dx^2} - 2 \frac{dy}{dx} + 2y = 0.$$

Show that $e^x \sin x$ also satisfies this equation.

*(34) Prove that, if $y \sqrt{1-x^2} = \sin^{-1} x$, then

$$(1-x^2) \frac{d^2 y}{dx^2} - 3x \frac{dy}{dx} - y = 0.$$

CHAPTER IV

MAXIMA AND MINIMA

§ 39. Critical Values.

We have seen that $\frac{dy}{dx}$ is the symbol for the rate of increase of y, regarded as a function of x, in terms of the variable x; and also that for any function which is capable of being represented by a graph, $\frac{dy}{dx} = \tan \psi$. It is easy from these facts to deduce the conditions under which the function y will attain any maximum or minimum values that it may possess, it being assumed that y and $\frac{dy}{dx}$ are both continuous functions of x.

For in the vicinity of a maximum value y will first be increasing, and then, after passing through the maximum value, will commence to decrease. Hence $\frac{dy}{dx}$ will first be positive and then negative, and hence, since $\frac{dy}{dx}$ is a continuous function, it must pass through the intermediate value, zero, at the actual value of x, where the maximum occurs (see § 12).

Similarly, near a minimum value, y will first be decreasing and then increasing. Then, using the same property of a continuous function, $\frac{dy}{dx}$ must be zero at the actual value of x where the minimum occurs.

It is often convenient to have a single term to denote the values of x for which y is either a maximum or a minimum, and we thus often speak of these values as the "critical" or "turning" values of the function. Our result may thus be expressed by saying that at critical values of either sort $\dfrac{dy}{dx}$ vanishes.

§ 40. Geometrical Investigation.

This is also evident geometrically, for if we draw the graph of the function we see that at a critical point the tangent is parallel to OX, and thus $\tan \psi = 0$.

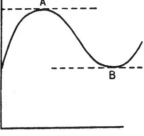

$$\therefore \quad \frac{dy}{dx} = 0.$$

We can thus state the first condition as follows:

The "critical values" of x (i.e. those values of x which

Fig. 6

make a function of x take either a maximum or minimum value) are included in the roots of the equation $\dfrac{dy}{dx} = 0$, it being supposed that the function and its derivative are continuous over the range of values considered.

We should especially note that an infinity does not therefore count as a maximum or minimum value, for, as we have previously shown, in the neighbourhood of an infinity the function is not continuous.

§ 41. Discrimination of Cases.

It remains for us to select which of these values of x give maxima, which give minima and which, if any, fail to give either. To deal with this question, we must consider the function as being represented by a graph or curve.

Let us consider the behaviour of the tangent to the curve in the neighbourhood of a maximum value of y. We define the positive direction of the tangent at a point as being the direction in which we look if we proceed along the curve from the origin in the direction in which the length of the curve is increasing. The angle ψ is then the angle between the positive direction of the tangent and the positive direction of OX.

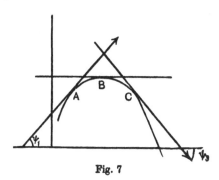

Fig. 7

When we are at A tan ψ_1 is positive. But when we get to C the angle ψ_3 is negative and tan ψ_3 is negative. Hence, during the interval from A to C, tan $\psi \left(= \dfrac{dy}{dx} \right)$ is decreasing.

But $\dfrac{d^2y}{dx^2}$ means the rate of increase of $\dfrac{dy}{dx}$, and hence in the neighbourhood of a maximum value $\dfrac{d^2y}{dx^2}$ is negative. This is certainly the case if $\dfrac{d^2y}{dx^2}$ is negative at the critical value. The only other possibility is that $\dfrac{d^2y}{dx^2}$ should be

negative near the critical
point, but zero at it, which
we consider later.

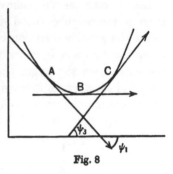

Fig. 8

   Similarly by examining
the figure in the neighbourhood of a minimum value

of y we see that here $\dfrac{dy}{dx}$

or $\tan \psi$ is increasing; hence

$\dfrac{d^2y}{dx^2}$ is positive in the neigh-

bourhood of a minimum value and, thus, at the minimum
value itself (the case mentioned being excepted as before).

§ 42. General Statement of Procedure.

We can thus state the elementary procedure for finding
maxima and minima as follows:

Obtain the roots of the equation $\dfrac{dy}{dx} = 0$.

Test at which of these roots $\dfrac{dy}{dx}$ changes from positive
to negative; or $\dfrac{d^2y}{dx^2}$ is negative. These values, when
substituted, give maximum values of y*.

Again, for minima, select those roots at which $\dfrac{dy}{dx}$
changes from negative to positive; or $\dfrac{d^2y}{dx^2}$ is positive.
These, when substituted, give minima.

---

  * In cases where the variable x is restricted in value by the conditions
of a particular problem, greatest or least values may also occur for the
"bounding" values of x. See worked Ex. (2), p. 54.

The cases in which $\dfrac{d^2y}{dx^2} = 0$ require further investigation and we must hold these over for the present.

§ 43. Worked Examples.

(1) *To find the maximum or minimum values of*

$$\frac{x}{x^2 + 1}.$$

Let $\qquad y = \dfrac{x}{x^2 + 1}.$

Then $\qquad \dfrac{dy}{dx} = \dfrac{1\,(x^2 + 1) - x\,(2x)}{(x^2 + 1)^2}$

$$= \frac{1 - x^2}{(x^2 + 1)^2}.$$

Hence $\dfrac{dy}{dx} = 0$ if $x^2 = 1$, and both y and $\dfrac{dy}{dx}$ are continuous functions.

Therefore the critical values are $x = \pm 1$.

Again

$$\frac{d^2y}{dx^2} = \frac{-2x\,(1 + x^2)^2 - 4x\,(1 + x^2)\,(1 - x^2)}{(1 + x^2)^4}$$

$$= \frac{-2x\,(1 + x^2) - 4x\,(1 - x^2)}{(1 + x^2)^3}.$$

When $\qquad x = +1, \;\; \dfrac{d^2y}{dx^2}$ is negative,

$$x = -1, \;\; \frac{d^2y}{dx^2} \text{ is positive.}$$

Thus $x = 1$ gives a maximum value $y = \frac{1}{2}$, and $x = -1$ gives a minimum value $y = -\frac{1}{2}$.

(2) The following example illustrates the elementary methods of dealing with an example involving two connected variables.

A window of perimeter 15 feet is formed by a rectangle surmounted by a semicircle. Find the dimensions which will admit a maximum amount of light.

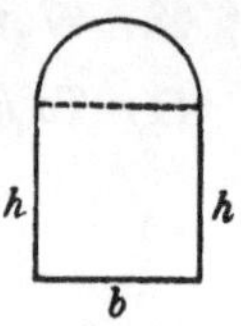

With the notation of the figure the area is

$$hb + \tfrac{1}{8}\pi b^2,$$

the perimeter $\quad 2h + b + \dfrac{\pi b}{2}.$

Fig. 9

We therefore have to make $bh + \tfrac{1}{8}\pi b^2$ a maximum, subject to the condition

$$2h + b + \frac{\pi b}{2} = 15.$$

We first get rid of h by substituting

$$h = \frac{1}{2}\left[15 - b\left(1 + \frac{\pi}{2}\right)\right].$$

$$\therefore \quad A = \frac{1}{2}\left[15b - b^2\left(1 + \frac{\pi}{2}\right)\right] + \frac{\pi b^2}{8},$$

$$\frac{dA}{db} = \frac{1}{2}\left[15 - 2b\left(1 + \frac{\pi}{2}\right)\right] + \frac{\pi b}{4}$$

$$= \frac{15}{2} - b - \frac{b\pi}{4}.$$

$$\therefore \quad \frac{dA}{db} = 0 \text{ if } \frac{15}{2} = b\left(1 + \frac{\pi}{4}\right),$$

i.e. if $\qquad b = \dfrac{30}{\pi + 4};$ whence $h = \dfrac{15}{\pi + 4}.$

It is easy to see that the value given is a maximum, since the minimum area clearly occurs when b is zero. In this, as in many practical cases, the variable b is not unrestricted in value. It must in fact lie between 0 and $\dfrac{30}{\pi+2}$. In such cases (see footnote, § 42) the bounding values of b must be considered, as well as the values obtained by equating the derivative to zero.

(3) Algebraic transformation may sometimes be profitably made before differentiating as follows.

Find maxima and minima of

$$\frac{(x-1)(x-6)}{x-10}.$$

Let $\qquad x-10=z.$

The expression now becomes

$$\frac{(z+9)(z+4)}{z},$$

or $\qquad z+13+\dfrac{36}{z}.$

$$\therefore f'(z)=1-\frac{36}{z^2},$$

and $\qquad f''z \;=\; +\dfrac{72}{z^3}.$

Critical values are $z=\pm 6$, and $+6$ gives a minimum value, and -6 a maximum value.

The minimum value is 25 and the maximum value 1. (See § 44.)

(4) It is also interesting to see how this example can be solved without the use of the Calculus.

Let $\qquad y=\dfrac{(x-1)(x-6)}{x-10}.$

$$\therefore \; xy-10y=x^2-7x+6,$$

i.e. $\qquad x^2-x(y+7)+10y+6=0.$

Now by supposition x is real, hence, by the Theory of Quadratics,

$$(y + 7)^2 \not< 40y + 24,$$

i.e. $$y^2 - 26y + 25 \not< 0,$$

i.e. $$(y - 25)(y - 1) \not< 0.$$

Hence y is either $\geqslant 25$ or $\leqslant 1$. Hence 25 is a minimum and 1 a maximum value of y.

§ 44. An Apparent Paradox.

The reader should notice, however, that in some cases the "commonsense" criterion is apt to be misleading. It is easily possible to have a function for which the maximum value (or one of the maximum values) is less than the minimum value (or one of the minimum values). The reason for this is seen in the restriction placed on the meaning of the words "maximum" and "minimum." For, in discussing the conditions, we only took into account the behaviour of the tangent immediately before and after a critical value, and there is nothing to prevent a value which gives a minimum in its own vicinity from being greater than a maximum in another part of the graph. To sum the matter up, throughout the discussion the words maximum and minimum are used in a purely localized sense.

This paradox is of quite frequent occurrence, particularly where an asymptote of the curve lies between critical values of x. Figs. 10 and 11 should show quite clearly how this apparent paradox may be explained. A case has already been encountered in the third example of § 43, where $z = 0$ is an asymptote.

Fig. 10 shows a case in which a minimum value of y occurring at D is greater than a maximum value of y

occurring at A in another part of the curve. Fig. 11 illustrates the same anomaly, in this case due to the existence of a value for which $y \to \infty$ between the two roots of $\frac{dy}{dx} = 0$. It should also be noticed that maxima and minima occur alternately, unless the curve is discontinuous. The student should construct various cases for himself and study their peculiarities.

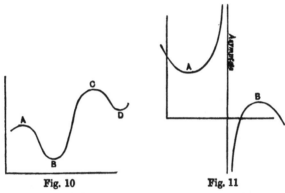

Fig. 10   Fig. 11

§ 45. Case of $\dfrac{d^2y}{dx^2} = 0$.

We may also consider a little further the case when

$$\frac{d^2y}{dx^2} = 0.$$

Now, as long as no further derivatives vanish, consider the graph of $\frac{dy}{dx}$ itself. If $\frac{d^2y}{dx^2} = 0$, the value under consideration will be a critical value on the graph of $\frac{dy}{dx}$. And since by hypothesis $\frac{dy}{dx}$ also is zero the graph of $\frac{dy}{dx}$ at the value considered will be as in one or other of Figs. 12 and 13. In this case then $\frac{dy}{dx}$ either increases up to zero and then decreases, or else decreases down to zero and then increases.

Remembering that $\frac{dy}{dx}$ means tan ψ, the corresponding portions of the original graph must be as in Figs. 14 and 15. Hence values of x which make $\frac{dy}{dx} = 0$ and also $\frac{d^2y}{dx^2} = 0$ do not necessarily give either maxima or minima

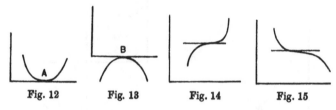

Fig. 12   Fig. 13   Fig. 14   Fig. 15

for y. They can give rise to the state of affairs illustrated in Figs. 14 and 15. These are known as "inflexional values" and "points of inflexion," and we shall consider these points again later.

§ 46. Maxima and Minima for two or more variables.

When two variables x and y are connected by a relation, it is possible to think of them as being both expressible in terms of some primary variable or parameter. Sometimes this parameter can be found easily, and sometimes it cannot, but in any case it exists and we can perform the differentiations required in terms of this instead of x. Similarly three variables connected by two relations can be conceived as all depending on a single parameter. For example, if

$y^2 - 4ax = 0$, we see that y and x can both be expressed in terms of t in the familiar form $x = at^2$, $y = 2at$. A method depending on this idea is given in the following section.

§ 47. The Method of Undetermined Multipliers.

Consider the following example.

Find the maximum or minimum values of u, given that

$$u = a^2 x^2 + b^2 y^2 + c^2 z^2,$$

$$x^2 + y^2 + z^2 = 1,$$

$$lx + my + nz = 0.$$

As we are dealing with three variables connected by two conditions, we can conceive x, y, z as being functions of some (undefined) parameter t.

Maxima and minima will then occur when $\dfrac{du}{dt} = 0$.

$$\therefore \quad \frac{du}{dt} = 2a^2 x \frac{dx}{dt} + 2b^2 y \frac{dy}{dt} + 2c^2 z \frac{dz}{dt} = 0.$$

Also since the other equations remain true for all values of x, y and z we can differentiate them with respect to t.

$$\therefore \quad 2x \frac{dx}{dt} + 2y \frac{dy}{dt} + 2z \frac{dz}{dt} = 0,$$

and
$$l \frac{dx}{dt} + m \frac{dy}{dt} + n \frac{dz}{dt} = 0.$$

We have therefore the three equations

$$\Sigma a^2 x \frac{dx}{dt} = 0, \quad \Sigma x \frac{dx}{dt} = 0, \quad \Sigma l \frac{dx}{dt} = 0,$$

to determine the critical values.

Now any two of these equations are sufficient to determine the ratios $\dfrac{dx}{dt} : \dfrac{dy}{dt}$ etc., and the third must therefore be redundant; i.e. must be compounded from the other two.

Therefore we must be able to find two constants A and B such that

$$\left. \begin{aligned} a^2x &= Ax + Bl \\ b^2y &= Ay + Bm \\ c^2z &= Az + Bn \end{aligned} \right\}.$$

The first equation will then be true by virtue of the other two.

Multiply these equations by x, y, z respectively and add. We get

$$a^2x^2 + b^2y^2 + c^2z^2 = A(x^2 + y^2 + z^2) + B(lx + my + nz).$$

$$\therefore \quad u = A, \text{ for } x^2 + y^2 + z^2 = 1, \; lx + my + nz = 0.$$

$$\therefore \quad \left. \begin{aligned} x(a^2 - u) &= Bl \\ y(b^2 - u) &= Bm \\ z(c^2 - u) &= Bn \end{aligned} \right\},$$

or

$$\left. \begin{aligned} \frac{x}{B} &= \frac{l}{a^2 - u} \\ \frac{y}{B} &= \frac{m}{b^2 - u} \\ \frac{z}{B} &= \frac{n}{c^2 - u} \end{aligned} \right\}.$$

Multiplying again by l, m, n respectively we get

$$\frac{l^2}{a^2 - u} + \frac{m^2}{b^2 - u} + \frac{n^2}{c^2 - u} = \frac{lx + my + nz}{B} = 0.$$

Now all the values of x, y, z have been those for which u is a maximum or minimum (for we used the equation

$\dfrac{du}{dt} = 0$). Hence the values of u given by this equation are the actual maxima and minima.

Hence the maximum and minimum values of u are the roots of the equation

$$\frac{l^2}{u-a^2} + \frac{m^2}{u-b^2} + \frac{n^2}{u-c^2} = 0.$$

The discrimination between maxima and minima in this method is difficult and rather outside the scope of this book. But this method can frequently be employed in Analytical Geometry, of both two and three dimensions.

EXAMPLES IV

(1) Divide a given number into two parts whose product is a maximum.

(2) Find the maximum and minimum values of $a \cos x + b \sin x$.

(3) Find the maximum and minimum values of

$$\frac{x^2 + 2x + 11}{x^2 + 4x + 10}.$$

(4) Investigate the critical values of $\dfrac{x}{2} + \cos x$.

(5) Find the dimensions of the greatest rectangle that can be inscribed in a given circle.

(6) A box is made by cutting a square from each of the corners of a rectangular piece of cardboard and folding up the sides. Find the dimensions of the box of maximum volume that can be so made from a sheet of cardboard $24'' \times 18''$.

(7) Find the cylinder of greatest volume that can be cut from a given solid sphere.

Find also the greatest right circular cone.

(8) Find the least area of sheet metal required to make an open cylindrical tank holding 20 cubic feet. (Neglect any overlapping.)

(9) An aperture in a tower consists of a rectangle with an equilateral triangle at either end of it. The total perimeter being 8 feet, find the dimensions giving a maximum area.

(10) On the ellipse $7x^2 + 3y^2 = 55$ the two points (2, 3), (1, 4) are taken. Find a third point on the curve which forms with these a triangle of maximum area; and prove that the tangent at this point is parallel to the chord joining the other two points.

*(11) Find the value of x for which the sum of the ordinates of the curves

$$y = 2x^3 - 15x^2 + 36x + 5,$$
$$y = x^3 - 4x + 3,$$

is a maximum, and show that at this value one of these ordinates is a maximum and the other a minimum.

*(12) Prove that the "turning-points" of the function $\dfrac{5x-13}{x^2-1}$ are $x = 5$ and $x = \frac{1}{5}$ and show that $\frac{1}{2}$ is a maximum and $\frac{25}{2}$ a minimum value of the function, explaining the apparent paradox.

*(13) If $f(x)$ denote $x - \dfrac{3 \sin x}{2 + \cos x}$ show that

$$f'(x) = \left(\frac{1 - \cos x}{2 + \cos x} \right)^2.$$

Show that $f'(x)$ attains its maximum value 4 when $\cos x = -1$, and give an approximate graph of $f(x)$ from $x = 0$ to $x = \pi$.

*(14) The superficial area of a right circular cylinder, including the two circular ends, is a given quantity S. If V is the volume show that

$$\frac{1}{V} \cdot \frac{dV}{dr} = \frac{l - 2r}{lr},$$

where l is the length, r the radius of the cylinder.

Show that the greatest volume is $(2\pi)^{-\frac{1}{2}} (\tfrac{1}{3}S)^{\frac{3}{2}}$ and prove that it is a true maximum.

*(15) Find the minimum value of $a \sec^2 \theta + b \csc^2 \theta$.

*(16) Find the values of θ that give maximum and minimum values of $\tan \theta + 3 \log \cos \theta + \theta$, if θ lies between 0 and $\dfrac{\pi}{2}$.

(17) With a given area A of sheet metal, find the ratio of base to height, if it is made into a cone of maximum volume (1) without a base, (2) with a base. (Neglect waste in cutting, etc.)

(18) Prove Rolle's Theorem, that if $f(x)$ is a rational algebraic function of x, one root at least of $f'(x) = 0$ lies between every consecutive pair of roots of $f(x) = 0$.

(19) Investigate the maxima and minima of $x^4 - 6x^2 + 8x$, and trace the curve $y = x^4 - 6x^2 + 8x$ between $x = -3$ and $x = 2$.

(20) Investigate critical values for the function

$$\frac{ax + b}{x + 1} - \frac{x}{(x + 1)^2}.$$

(21) Test for maxima and minima the function $\dfrac{4}{x} + \dfrac{1}{1 - x}$ and show that no values of the function lie between 1 and 9.

(22) A rectangular box whose end is a square is tied by a cord passing round the box once lengthways, and twice round its girth. Find the maximum value of the volume in terms of the length of the string (neglecting knots, etc.).

(23) Find the critical values of (a) xy, (b) $x^2 + y^2$, given that

$$\frac{x}{a} + \frac{y}{b} = 1.$$

Interpret these results geometrically.

(24) Use the method of undetermined multipliers to prove that the lengths of the maximum and minimum semi-diameters of the conic $ax^2 + 2hxy + by^2 = 1$ are the roots of the equation

$$\left(a - \frac{1}{r^2}\right)\left(b - \frac{1}{r^2}\right) = h^2.$$

N.B. This gives a method for finding the axes of the general conic.

(25) Find the length and position of the least focal chord of

$$y^2 = 4ax.$$

(26) Show that the shortest line which may be drawn to a parabola from a point on its axis is a normal to the curve.

CHAPTER V

APPLICATION TO GEOMETRY OF CURVES

§ 48. Application to Geometry.

We have seen in § 19 that $\frac{dy}{dx} = \tan \psi$, where ψ is the inclination to the axis of x, of the tangent to the curve whose equation in rectangular Cartesian coordinates is $y = y(x)$.

It thus follows that we can always write down the equations to tangents and normals of any curves whose equations are known. We shall give in this chapter a brief account of the more elementary methods connected with the Geometry of Curves, in regard to both Cartesian and Polar Coordinates*.

§ 49. Equation of Tangent and Normal.

To avoid confusion we will denote the point of contact by (x_0, y_0) and the value of $\frac{dy}{dx}$ at that point (i.e. first calculating $\frac{dy}{dx}$ from the equation and then substituting) by $\left(\frac{dy}{dx}\right)_0$. Other suffix notation used later will bear a similar meaning.

The equation of the tangent will now be

$$(y - y_0) = \left(\frac{dy}{dx}\right)_0 \{x - x_0\},$$

---

* It is assumed in all cases of Cartesian coordinates that equal scales of length are employed on the two axes. Otherwise our results have to be modified in some cases.

for this represents the straight line through (x_0, y_0) at the given inclination. Similarly the equation to the normal, which is a line through (x_0, y_0) perpendicular to the tangent, must be

$$\left(\frac{dy}{dx}\right)_0 (y - y_0) + (x - x_0) = 0.$$

E.g. Find the equation to the tangent and normal to $y = x^2 + x + 1$ at the point where $x = 2$.

If $x = 2$, $y = 7$:

also $\dfrac{dy}{dx} = 2x + 1$

$= 5$ if $x = 2$

Hence equation to tangent is $(y - 7) = 5(x - 2)$,

equation to normal is $5(y - 7) + x - 2 = 0$.

§ 50. Two Modifications of this form.

Two modifications frequently arise.

(1) Sometimes the equation to the curve is not given explicitly in the form $y = f(x)$ but implicitly, as for example $x^3 + y^3 - 3axy = 0$. In this case we must use the formula

$$\frac{dy}{dx} = -\frac{\dfrac{\partial u}{\partial x}}{\dfrac{\partial u}{\partial y}} \qquad (\S 38).$$

The equation of a tangent may now be written

$$(x - x_0)\left(\frac{\partial u}{\partial x}\right)_0 + (y - y_0)\left(\frac{\partial u}{\partial y}\right)_0 = 0.$$

Similarly the equation to the normal becomes

$$(x - x_0)\left(\frac{\partial u}{\partial y}\right)_0 - (y - y_0)\left(\frac{\partial u}{\partial x}\right)_0 = 0,$$

or
$$\frac{x - x_0}{\left(\dfrac{\partial u}{\partial x}\right)_0} = \frac{y - y_0}{\left(\dfrac{\partial u}{\partial y}\right)_0}.$$

Thus in the curve $x^3 + y^3 - 3axy = 0,$

$$\left(\frac{\partial u}{\partial x}\right)_0 = 3\,(x_0{}^2 - ay_0), \quad \left(\frac{\partial u}{\partial y}\right)_0 = 3\,(y_0{}^2 - ax_0).$$

Hence the equation to the tangent is

$$(x - x_0)\,(x_0{}^2 - ay_0) + (y - y_0)\,(y_0{}^2 - ax_0) = 0,$$

which reduces to the form

$$x\,(x_0{}^2 - ay_0) + y\,(y_0{}^2 - ax_0) - ax_0y_0 = 0,$$

remembering that (x_0, y_0) itself lies on the curve.

Similarly the normal may be written down.

(2) Very often the best method in which the equation to a curve can be expressed is in parametric notation: that is, x and y are both given in terms of some third variable t.

In such cases we find $\dfrac{dy}{dx}$ from the formula

$$\frac{dy}{dx} = \frac{\dfrac{dy}{dt}}{\dfrac{dx}{dt}}.$$

The equation to the tangent now becomes

$$(y - y_0)\left(\frac{dx}{dt}\right)_0 = (x - x_0)\left(\frac{dy}{dt}\right)_0,$$

and a corresponding form

$$(y - y_0)\left(\frac{dy}{dt}\right)_0 + (x - x_0)\left(\frac{dx}{dt}\right)_0 = 0$$

is seen to be the equation to the normal.

On no account should the student try to get the Cartesian equation of the curve by eliminating the parameter. It is nearly always by reason of the difficulty of the Cartesian equation that this parametric mode of representation has been resorted to, and any attempt to eliminate the parameter almost invariably increases the difficulties.

Thus if $x = t^2$, $y = 4t - t^3$ is the equation

$$\frac{dy}{dx} = \frac{4 - 3t^2}{2t}.$$

Hence the equation to the tangent is

$$2t\,(y - 4t + t^3) = (4 - 3t^2)\,(x - t^2),$$

i.e. $\quad 2ty - (4 - 3t^2)\,x = t^2\,(t^2 + 4),$

and similarly for the normal.

§ 51. Arc-Derivatives.

In Fig. 16, P and P' are close points on a curve whose equation is given, and we have already seen that

$$\frac{P'L}{PL} = \frac{\delta y}{\delta x} = \tan P'PL.$$

Now in the limit, when P and P' become very close, the arc PP' cannot be distinguished from the chord PP' (that is, of course, provided there is no irregularity of the curve near P), and therefore we may take

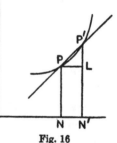

Fig. 16

the length of the chord PP' as representing δs, "the element of the arc of the curve."

Then $\qquad \cos P'PL = \dfrac{PL}{PP'} = \dfrac{\delta x}{\delta s},$

and $\quad \sin P'PL = \dfrac{P'L}{PP'} = \dfrac{\delta y}{\delta s}.$

But in the limit the angle $P'PL$ becomes the angle which we denote by ψ, and we arrive thus at the two formulae

$$\frac{dx}{ds} = \cos \psi, \quad \frac{dy}{ds} = \sin \psi.$$

We also have $\left(\dfrac{dx}{ds}\right)^2 + \left(\dfrac{dy}{ds}\right)^2 = 1,$

and multiplying both sides by $\left(\dfrac{ds}{dt}\right)^2$ we arrive at

$$\left(\frac{ds}{dt}\right)^2 = \left(\frac{dx}{dt}\right)^2 + \left(\frac{dy}{dt}\right)^2.$$

§ 52. Subtangents and Subnormals.

All lengths in the figure can now be calculated by the ordinary methods of Trigonometry and Coordinate Geometry. But due regard must be had to signs, remembering that ψ is the angle between the positive direction of the tangent and the axis, or mistakes will soon occur.

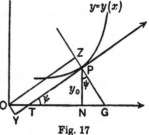

Fig. 17

Thus in Fig. 17 the "tangent" $PT = y_0 \operatorname{cosec} \psi$, the "normal" $PG = y_0 \sec \psi$, the "subtangent" $TN = y_0 \cot \psi$, and the "subnormal" $NG = y_0 \tan \psi$.

Also OY the perpendicular from O on the tangent*

$$= \pm \frac{y_0 - x_0 \left(\dfrac{dy}{dx}\right)_0}{\sqrt{1 + \left(\dfrac{dy}{dx}\right)_0^2}}.$$

This perpendicular is usually denoted by p.

* Use the equation of § 49.

§ 53. Worked Example.

Prove that the perpendicular on the tangent at (x, y) is given by $p = x \sin \psi - y \cos \psi$ and that $\frac{dp}{d\psi}$ represents the perpendicular from O on the corresponding normal.

In the figure $\qquad OY = OT \sin \psi$,

and $\qquad\qquad\qquad OT = ON - NT$

$$= x_0 - y_0 \cot \psi.$$

$$\therefore \quad OY = x_0 \sin \psi - y_0 \cos \psi,$$

or, dropping the suffixes, we have that at (x, y)

$$p = x \sin \psi - y \cos \psi.$$

Hence

$$\frac{dp}{d\psi} = \frac{dx}{d\psi} \sin \psi + x \cos \psi$$

$$- \frac{dy}{d\psi} \cos \psi + y \sin \psi$$

$$= x \cos \psi + y \sin \psi + \frac{ds}{d\psi} \left\{ \frac{dx}{ds} \sin \psi - \frac{dy}{ds} \cos \psi \right\}.$$

But since

$$\frac{dx}{ds} = \cos \psi \quad \text{and} \quad \frac{dy}{ds} = \sin \psi,$$

$$\frac{dx}{ds} \sin \psi = \frac{dy}{ds} \cos \psi.$$

Hence

$$\frac{dp}{d\psi} = x \cos \psi + y \sin \psi.$$

Now

$$OZ = PY$$

$$= PT + TY$$

$$= y \operatorname{cosec} \psi + OY \cot \psi$$

$$= y \operatorname{cosec} \psi + \cot \psi \{x \sin \psi - y \cos \psi\}$$

$$= x \cos \psi + y \frac{1 - \cos^2 \psi}{\sin \psi}$$

$$= x \cos \psi + y \sin \psi = \frac{dp}{d\psi}.$$

EXAMPLES V A

Write down the equations to the tangents and normals to the following curves at the points specified:

(1) $y = x^3 + 4x + 3$, where $x = 2$.

(2) $y = x^2(x+3)$, at the point where it *cuts* the axis of x.

(3) $x^2 + y^2 = 25$, where $x = 3$.　　(4) $y^2 = 4ax$, where $x = a$.

(5) $x = a \cos \theta$, $y = b \sin \theta$, where $\theta = \dfrac{\pi}{4}$.

*(6) $2y = 2x^3 - 4x^2 - 7x + 3$, where it meets the axis of y.

(7) $\left.\begin{array}{l} x = a\,(t + \sin t) \\ y = a\,(1 + \cos t) \end{array}\right\}$ (the cycloid), at any point "t."

(8) $x^3 + y^3 - 3xy = 0$, at (x_0, y_0).

(9) $ax^2 + 2hxy + by^2 = 1$, at (x_0, y_0).　　(10) $xy = c^2$, at (x_0, y_0).

(11) The parabola $x = at^2$, $y = 2at$ at any point.

(12) The "Catenary" $y = \dfrac{c}{2}\left(e^{\frac{x}{c}} + e^{-\frac{x}{c}}\right)$ at any point.

*(13) Show that the coordinates of any point on $y^2 = x^2(1-x)$ can be expressed in terms of a parameter by the equations

$$x = \frac{y}{t} = 1 - t^2,$$

and find the equation to the tangent at the point of parameter t_0.

Prove that the tangents at two points of parameter t_0 and $\frac{1}{t_0}$ meet on the curve itself.

*(14) Find the equation of the tangent at any point of the "semi-cubical parabola" $ay^2 = x^3$.

The tangent at a point Q of the semi-cubical parabola meets the curve again in R and QR is divided in the ratio $k : 1$.

Show that the locus of the point of division is another semi-cubical parabola, and find the value of k for which the locus becomes the axis of x.

*(15) Find the equation to the tangent to the curve $3ay^2 = x^2(x+a)$ at any point. The tangent is drawn to this curve at $(2a, 2a)$.

Find the point where it meets the curve again and show that it is the normal at that point.

(16) Write down the equation of the tangent at any point of the curve $2x^3 = 3(x^2 + y^2)$.

(17) Find the equation to the tangent at any point of the curve given by

$$x = \frac{am}{1+m^3}, \quad y = \frac{am^2}{1+m^3}.$$

By choosing a proper value of m, deduce the equation to the asymptote.

(18) A curve is defined by the equations

$$\left. \begin{array}{l} x = a\sin^2\theta\cos\theta \\ y = a\sin^3\theta - 2a\sin\theta \end{array} \right\}.$$

Find the expressions for the subtangent and subnormal at any point.

Also find the length of the perpendicular from O on any tangent.

(19) Find $\frac{dx}{ds}$ and $\frac{dy}{ds}$ for the curve $y = c\log\sec\frac{x}{c}$.

(20) Show that the line $\dfrac{x}{a} + \dfrac{y}{b} = 2$ is a tangent to the curve

$$\left(\frac{x}{a}\right)^n + \left(\frac{y}{b}\right)^n = 2.$$

(21) What is the condition that the curves

$$ax^2 + by^2 = 1, \quad a'x^2 + b'y^2 = 1,$$

should cut orthogonally?

(22) Prove that the length of the portion of the tangent to

$$x^{\frac{2}{3}} + y^{\frac{2}{3}} = c^{\frac{2}{3}},$$

intercepted between the axes is constant.

(23) Find the subtangent to the curve $xy^2 = a^2(a-x)$.

For what point is this a maximum?

(24) In the cycloid $x = a(\theta + \sin\theta)$, $y = a(1 - \cos\theta)$, prove that $\psi = \dfrac{\theta}{2}$ and find an expression for p.

(25) Prove that the two parabolas $y^2 = ax$, $x^2 = ay$ meet on the Folium $x^3 + y^3 = 2axy$.

At what angles do these curves cut one another?

*(26) The tangent at a point "t" of the cycloid of Ex. 7 is the normal to the cycloid at the point where it meets the next span on the right. Prove that $\cot\dfrac{t}{2} = \dfrac{\pi}{2}$.

(27) Prove that any point on $ay^2 = x^3$ may be represented by the "parametric notation" $x = at^2$, $y = at^3$. Show that if P and Q are two points, such that the chord PQ always passes through a fixed point on the curve, the tangents at P and Q will meet on a parabola.

§ 54. The "Class" of a Curve.

Taking the equation to a tangent in the form

$$(x - x_0)\left(\frac{\partial u}{\partial x}\right)_0 + (y - y_0)\left(\frac{\partial u}{\partial y}\right)_0 = 0,$$

we see that the condition that the tangent at some point (x_0, y_0) should pass through (h, k) is

$$(h - x_0) \left(\frac{\partial u}{\partial x}\right)_0 + (k - y_0) \left(\frac{\partial u}{\partial y}\right)_0 = 0.$$

This equation* therefore represents a curve which goes through all the points of contact of the tangents drawn from (h, k) to the curve, and the intersections of this with the curve are the points of contact.

The curve $(h - x) \left(\frac{\partial u}{\partial x}\right) + (k - y) \left(\frac{\partial u}{\partial y}\right) = 0$ is apparently of degree n, but it will be found that, by using the equation to the original curve, it may always be reduced to a curve of degree $(n - 1)$. Hence there are $n (n - 1)$ intersections and thus $n (n - 1)$ tangents can in general be drawn from an external point to a curve of degree n.

The number of tangents that can be drawn from an external point to a curve is known as the "class" of the curve. Hence *the class of a curve of degree n is $n (n - 1)$*.

Similarly the normal at (x, y) goes through (h, k) if

$$(h - x) \left(\frac{\partial u}{\partial y}\right) - (k - y) \left(\frac{\partial u}{\partial x}\right) = 0.$$

This*, again, is a curve, this time of the nth degree, passing through the feet of all the normals that can be drawn from (h, k) to the curve. This meets the curve in n^2 points and hence *in general n^2 normals can be drawn to a curve of degree n from any point.*

§ 55. Example.

The above results are well illustrated by the ellipse

$$\frac{x^2}{a^2} + \frac{y^2}{b^2} = 1.$$

---

* The suffixes may now be removed, for we are now fixing our attention on the points of contact.

For since $\quad \dfrac{\partial u}{\partial x} = \dfrac{2x}{a^2}, \quad \dfrac{\partial u}{\partial y} = \dfrac{2y}{b^2},$

the points of contact of the tangents from (h, k) lie on

$$(h - x)\frac{2x}{a^2} - (k - y)\frac{2y}{b^2} = 0,$$

i.e. $\qquad\qquad \dfrac{hx}{a^2} + \dfrac{ky}{b^2} = \dfrac{x^2}{a^2} + \dfrac{y^2}{b^2} = 1.$

(Note that the equation reduces to one of the first degree.)

Similarly the feet of the normals lie on the curve

$$(h - x)\frac{y}{b^2} - (k - y)\frac{x}{a^2} = 0,$$

which is a rectangular hyperbola. Thus four normals can be drawn from (h, k) to the ellipse, and their feet lie on a rectangular hyperbola whose equation is

$$(h - x)\frac{y}{b^2} - (k - y)\frac{x}{a^2} = 0.$$

If we put $(a \cos \theta, b \sin \theta)$ for (x, y) we get the usual equation in θ given in the text-books on Conics.

§ 56. Polar Coordinates.

Let P and P' be two close points on a curve whose equation is given in polar coordinates. Then P is (r, θ) and P' is $(r + \delta r, \theta + \delta\theta)$, and $\angle P'OP = \delta\theta$ which is small. Now draw PN perpendicular to OP'.

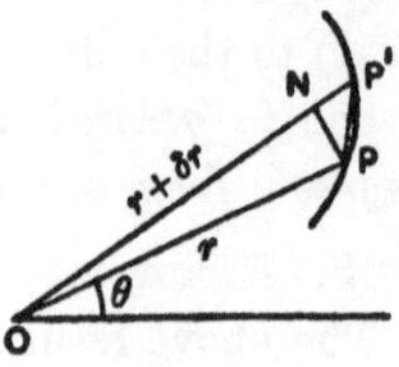

Fig. 18

$$\therefore \quad PN = r \sin \delta\theta = r\delta\theta,$$

$$P'N = OP' - ON = r + \delta r - r \cos \delta\theta$$

$$= \delta r,$$

since $\cos \delta\theta \to 1.$

Again, in the limit the arc $PP' =$ chord PP',

i.e. $$PP' = \delta s \,;$$

$$\therefore \quad \left. \begin{aligned} r\,\frac{\delta\theta}{\delta r} &= \tan PP'N \\ r\,\frac{\delta\theta}{\delta s} &= \sin PP'N \\ \frac{\delta r}{\delta s} &= \cos PP'N \end{aligned} \right\}.$$

Now in the limit when P and P' coincide the angle $PP'N$ becomes the angle between the tangent at P and the radius vector OP. This angle is always denoted by ϕ, and we thus have the formulae:

$$\left. \begin{aligned} \tan\phi &= r\,\frac{d\theta}{dr} \\ \sin\phi &= r\,\frac{d\theta}{ds} \\ \cos\phi &= \frac{dr}{ds} \end{aligned} \right\},$$

and $$r^2\left(\frac{d\theta}{dr}\right)^2 + \left(\frac{dr}{ds}\right)^2 = 1.$$

§ 57. Alternative proof.

The student who is acquainted with de Moivre's Theorem should notice the following neat proof of these results.

It is obvious from a figure that $\psi = \phi + \theta$. Now

$$\begin{aligned} x + iy &= r\cos\theta + ir\sin\theta \\ &= r\left(\cos\theta + i\sin\theta\right) \\ &= re^{i\theta}. \end{aligned}$$

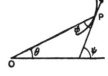

Fig. 19

Differentiate both sides of this relation with respect to s.

$$\therefore \quad \frac{dx}{ds} + i\frac{dy}{ds} = \frac{dr}{ds} \cdot e^{i\theta} + ire^{i\theta} \cdot \frac{d\theta}{ds},$$

$$\therefore \quad \cos\psi + i\sin\psi = e^{i\theta}\left\{\frac{dr}{ds} + ir\frac{d\theta}{ds}\right\},$$

i.e.
$$e^{i\psi} = e^{i\theta}\left\{\frac{dr}{ds} + ir\frac{d\theta}{ds}\right\},$$

i.e.
$$\frac{dr}{ds} + ir\frac{d\theta}{ds} = e^{i(\psi-\theta)} = e^{i\phi}$$
$$= \cos\phi + i\sin\phi.$$

Hence, equating real and imaginary parts,

$$\frac{dr}{ds} = \cos\phi, \quad r\frac{d\theta}{ds} = \sin\phi,$$

and dividing,

$$\tan\phi = \frac{r\dfrac{d\theta}{ds}}{\dfrac{dr}{ds}} = r\frac{d\theta}{dr}.$$

We can also easily find p (the perpendicular from O on any tangent).

For
$$\begin{aligned} p &= r\sin\phi \\ &= r\sin(\psi-\theta) \\ &= r\sin\psi\cos\theta - r\cos\psi\sin\theta \\ &= x\sin\psi - y\cos\psi. \end{aligned}$$

§ 58. Polar Subtangent and Subnormal.

By remembering these results and drawing the figure, we can obtain all the information we require for curves in Polar coordinates. Thus if tOT is drawn as in Fig. 20 perpendicular to OP and meeting the tangent in T and the normal in t, the length OT is known as the Polar Subtangent and is clearly

$$r \tan \phi = r^2 \frac{d\theta}{dr}.$$

Similarly Ot is known as the Polar Subnormal and is clearly

$$r \cot \phi = \frac{dr}{d\theta}.$$

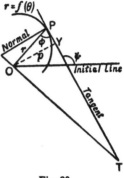

Fig. 20

In the same way the length and inclination of the perpendicular OY can be calculated and hence the equation to the tangent may be written down in the form $r \cos (\theta - \alpha) = p$. Similarly the perpendicular on the normal is equal to PY or $r \cos \phi$, and the inclination of this perpendicular is ψ, so that the equation to the normal may also be written down.

§ 59. Worked Example.

Find the equation to the tangent and normal at any point of the cardioid $r = a (1 + \cos \theta)$.

Let P be (r_1, θ_1).

We have
$$\tan \phi = r_1 \left(\frac{d\theta}{dr}\right)_1 = \frac{r_1}{\left(\dfrac{dr}{d\theta}\right)_1}$$

$$= \frac{a (1 + \cos \theta_1)}{- a \sin \theta_1} = - \cot \frac{\theta_1}{2}.$$

$$\therefore \quad \phi = \frac{\pi}{2} + \frac{\theta_1}{2} \text{ and } \psi = \phi + \theta_1 = \frac{\pi}{2} + \frac{3\theta_1}{2}.$$

Also
$$p = r \sin \phi = a (1 + \cos \theta_1) \cos \frac{\theta_1}{2}.$$

The inclination of the perpendicular is $\frac{3\theta_1}{2}\left(\text{i.e. } \psi - \frac{\pi}{2}\right)$.

Hence the Cartesian equation of tangent is

$$x \cos \frac{3\theta_1}{2} + y \sin \frac{3\theta_1}{2} = a\left(1 + \cos \theta_1\right) \cos \frac{\theta_1}{2},$$

or $\quad r \cos \left(\theta - \frac{3\theta_1}{2}\right) = a\left(1 + \cos \theta_1\right) \cos \frac{\theta_1}{2} = 2a \cos^3 \frac{\theta_1}{2}.$

Similarly the perpendicular on the normal

$$= r_1 \sin \frac{\theta_1}{2} = a\left(1 + \cos \theta_1\right) \sin \frac{\theta_1}{2};$$

and the inclination of this perpendicular is ψ, i.e. $\frac{\pi}{2} + \frac{3\theta_1}{2}$.

Hence the equation of normal is

$$x \sin \frac{3\theta_1}{2} - y \cos \frac{3\theta_1}{2} = a\left(1 + \cos \theta_1\right) \sin \frac{\theta_1}{2},$$

or in polar coordinates

$$r \sin \left(\frac{3\theta_1}{2} - \theta\right) = 2a \cos^2 \frac{\theta_1}{2} \sin \frac{\theta_1}{2}.$$

EXAMPLES V B

(1) Tangents are drawn from a given point to a curve

$$Ax^3 + By^2 = 1;$$

find the equation of the conic on which the points of contact of these tangents lie.

*(2) Prove that three tangents only can be drawn from the origin to the curve $y = x^3 - 13x^2 + 10x - 36$ and find their points of contact.

(3) Tangents are drawn from (h, k) to the curve $ay^2 = x^3$. Find the equation of the curve on which the points of contact of these tangents lie.

(4) Prove that a radius vector through O at an angle of 30° with the axis of x is a normal to the curve $(x^2 + 9y^2) x = 24y$ at both its extremities.

*(5) If in the equation of any curve we put $\dfrac{1}{r} = u$ prove the formula

$$\dfrac{1}{p^2} = u^2 + \left(\dfrac{du}{d\theta}\right)^2.$$

(6) Find ϕ and p in the spiral $r = ae^{\theta \cot \alpha}$.

(7) In the parabola $\dfrac{2a}{r} = 1 - \cos\theta$, show that $p^2 = ar$, and that the polar subtangent $= 2a \operatorname{cosec}\theta$.

(8) Find the equation connecting p and r for the cardioid

$$r = a(1 + \cos\theta).$$

*(9) In the rectangular hyperbola $r^2 \cos 2\theta = a^2$, prove that the perpendicular from O on any tangent makes with the initial line the same angle as the radius vector of the point of contact.

(10) With the usual notation, prove that in the family of curves given by the equation $r^n = a^n \cos n\theta$, p and r are connected by the relation $r^{n+1} = a^n p$.

(11) If two tangents to the cardioid $r = a(1 + \cos\theta)$ are parallel, show that the line joining their points of contact subtends an angle $\dfrac{2\pi}{3}$ at the cusp.

(12) Find the perpendicular on the tangent at the point of the Limaçon $r = a + b\cos\theta$, where $\theta = \dfrac{\pi}{2}$.

(13) Find an expression for the polar subnormal of the spiral $r = ae^{\kappa\theta}$, and prove that the locus of its extremity is another spiral.

(14) Find the equation to the tangent to the curve $r^2 \cos 2\theta = a^2$, where $\theta = a$.

(15) Find the equation to the tangent at any point on the Lemniscate $r^2 = a^2 \cos 2\theta$, and find an expression for p.

(16) In the usual notation, prove the formula

$$r^2 = p^2 + \left(\dfrac{dp}{d\psi}\right)^2.$$

CHAPTER VI

THE EXPANSION OF A FUNCTION AS AN INFINITE SERIES

§ 60. Successive Differentiation.

It is occasionally necessary to write down the nth derivative of a given function. Only in very special cases can any formula be found for the nth derivative and the methods used are largely inductive. Some of the commonest are indicated in the following three worked examples.

(1) *Find the nth derivative of $\dfrac{1}{x-a}$.*

Here
$$y = \frac{1}{x-a} \qquad = (x-a)^{-1},$$
$$y_1 = \qquad -1\,(x-a)^{-2},$$
$$y_2 = \qquad 1.2\,(x-a)^{-3},$$
$$y_3 = \qquad -1.2.3\,(x-a)^{-4}.$$

Proceeding in this way it is clear that we shall finally get
$$y_n = (-1)^r\, n!\,(x-a)^{-(n+1)},$$
or
$$(-1)^n\, \frac{n!}{(x-a)^{n+1}}.$$

(2) *Find the nth derivative of $\sin x$.*

Here
$$y = \sin x,$$
$$y_1 = \cos x = \sin\left(\frac{\pi}{2} + x\right).$$
$$\therefore \quad y_2 = \cos\left(\frac{\pi}{2} + x\right) = \sin\left(\pi + x\right),$$

and so on.

Clearly $$y_n = \sin\left(x + \frac{n\pi}{2}\right).$$

(3) *Find the nth derivative of $e^{ax}\sin bx$.*

(This is a form frequently encountered, and the result should be remembered.)

$$y = e^{ax}\sin bx,$$
$$y_1 = e^{ax}\{a\sin bx + b\cos bx\}$$
$$= \sqrt{a^2+b^2}\, e^{ax}\left\{\frac{a}{\sqrt{a^2+b^2}}\sin bx + \frac{b}{\sqrt{a^2+b^2}}\cos bx\right\}.$$

Now let $$\tan^{-1}\frac{b}{a} = \alpha,$$

then $$\frac{b}{\sqrt{a^2+b^2}} = \sin\alpha, \quad \text{and} \quad \frac{a}{\sqrt{a^2+b^2}} = \cos\alpha.$$

$$\therefore \quad y_1 = \sqrt{a^2+b^2}\, e^{ax}\sin(bx+\alpha).$$

Thus a single differentiation preserves the form of the expression but multiplies it by $\sqrt{a^2+b^2}$ and adds the constant α to the angle. Hence n differentiations will repeat this process n times and thus

$$y_n = (a^2+b^2)^{\frac{n}{2}} e^{ax}\sin(bx+n\alpha).$$

A precisely similar result is true for $e^{ax}\cos bx$.

§ 61. Leibnitz' Theorem.

There is one general theorem on this subject. It is known as Leibnitz' Theorem on Successive Differentiation and is the extension of the rule for the differentiation of a product. If $y = uv$ and differentiations are denoted by suffixes we see that

$$y_1 = uv_1 + u_1v,$$
$$y_2 = uv_2 + u_1v_1 + u_1v_1 + u_2v^*$$
$$= uv_2 + 2u_1v_1 + u_2v.$$

Similarly $$y_3 = uv_3 + 3u_1v_2 + 3u_2v_1 + u_3v.$$

---

* Treating uv_1 and u_1v themselves as products.

The analogy between these coefficients and those of the Binomial expansion suggests as the result

$$y_n = uv_n + nu_1v_{n-1} + {}^nC_2u_2v_{n-2} + \ldots + {}^nC_ru_rv_{n-r} + \ldots + u_nv.$$

This result can easily be proved by induction, by assuming the relation true for some value of n and differentiating it again. Then by using the identity

$$^{n+1}C_r = {}^nC_r + {}^nC_{r-1},$$

we get the same form with $n + 1$ in place of n.

The proof is left to the reader as an exercise.

§ 62. Expansions in Infinite Series.

The Calculus is of great use in obtaining expansions of given functions in ascending powers of x, where such expressions exist. Care must, of course, be taken in regard to the convergency of the series employed, and also of the results obtained, otherwise serious errors may result. Methods which will suffice for many simple series are given in the following articles. We shall assume as having been proved elsewhere the Binomial Theorem for any index, and the Exponential series. We therefore assume the following expansions

$$(1 + x)^n = 1 + nx + \frac{n(n-1)}{2!}x^2 + \frac{n(n-1)(n-2)}{3!}x^3 + \ldots,$$

for all values of n provided $|x| < 1$,

$$\text{and} \quad e^x = 1 + x + \frac{x^2}{2!} + \frac{x^3}{3!} + \ldots \text{ ad inf.}$$

for all values of x.

§ 63. The Logarithmic Series.

To find a series for $\log_e(1 + x)$.

Let $\quad \log_e(1 + x) \equiv a_0 + a_1x + a_2x^2 + \ldots + a_nx^n + \ldots.$

Differentiate both sides of this identity.

$$\therefore \quad \frac{1}{1+x} \equiv a_1 + 2a_2x + 3a_3x^2 + \dots \text{ ad inf.}$$

But, by the Binomial Theorem,

$$\frac{1}{1+x} = (1+x)^{-1} \equiv 1 - x + x^2 - x^3 + \dots \text{ ad inf.,}$$

provided that x lies between $+1$ and -1.

Hence $\qquad a_1 = 1, \quad a_2 = -\frac{1}{2}, \quad a_3 = \frac{1}{3},$

and so on.

Also putting $x = 0$ in the original series, $a_0 = \log 1 = 0$.

Hence the series (which is only valid if x lies between $+1$ and -1) is given by

$$\log(1+x) \equiv x - \frac{x^2}{2} + \frac{x^3}{3} - \dots + (-1)^{n-1}\frac{x^n}{n} + \dots \text{ ad inf.}$$

Changing x into $-x$ we get also the series

$$\log(1-x) \equiv -x - \frac{x^2}{2} - \frac{x^3}{3} - \dots \text{ ad inf.}$$

§ 64. Calculation of Logarithms.

A modification of these series is used to calculate natural logarithms. It is possible to use the series themselves; for example, by putting $x = \frac{1}{2}$ in the series for $\log(1-x)$ we get an expression for $\log_e 2$; but the series does not converge very rapidly, and a very large number of terms must be taken in order to obtain a high degree of accuracy. To obtain a series which will converge more rapidly (and thus necessitate less calculation) we proceed as follows.

Subtract the series of the last article from one another. We get the series

$$\log_e \frac{1+x}{1-x} \equiv 2\left\{x + \frac{x^3}{3} + \frac{x^5}{5} + \dots\right\}.$$

Now put $x = \dfrac{p-q}{p+q}$ and the series reduces to

$$\log_e \frac{p}{q} = 2\left\{\left(\frac{p-q}{p+q}\right) + \frac{1}{3}\left(\frac{p-q}{p+q}\right)^3 + \frac{1}{5}\left(\frac{p-q}{p+q}\right)^5 + ...\right\}.$$

If p and q are chosen very nearly equal this series converges very rapidly and only the first two or three terms are required.

Thus to find $\log_e 2$. Let $p=2$, $q=1$.

$$\therefore \quad \log_e 2 = 2\left[\frac{1}{3} + \frac{1}{3}\cdot\frac{1}{3^3} + \frac{1}{5}\cdot\frac{1}{3^5} + ...\right]$$

$= \cdot6666\,66$	$\dfrac{2}{3} = \cdot666666$	
$\cdot0246\,91$	$\dfrac{2}{3^3} = \cdot074074$	$\div 3$
$\cdot0016\,46$	$\dfrac{2}{3^5} = \cdot008230$	$\div 5$
$\cdot0001\,30$	$\dfrac{2}{3^7} = \cdot000914$	$\div 7$
$\cdot0000\,12$	$\dfrac{2}{3^9} = \cdot000102$	$\div 9$
$\cdot6931$	correct to four places of decimals.	

Similarly putting $p=3$, $q=2$ we can find $\log_e 3 - \log_e 2$ and hence $\log_e 3$ and so on. The above convenient arrangement of the numerical work should be noted.

§ 65. Expansions of sin x and cos x.

Let $\sin x$ (radians) $\equiv a_0 + a_1 x + a_2 x^2 + a_3 x^3 + ...$

Now since $\sin x$ changes sign when x changes sign, we see that the series can only comprise odd powers of x.

Hence $a_0 = a_2 = a_4$, etc. $= 0$.

$$\therefore \quad \sin x \equiv a_1 x + a_3 x^3 + a_5 x^5 + ...$$

Differentiating both sides

$$\cos x \equiv a_1 + 3a_3 x^2 + 5a_5 x^4 + \dots.$$

Differentiating again

$$-\sin x \equiv 2 \cdot 3a_3 x + 4 \cdot 5a_5 x^3 + \dots.$$

Comparing the two series for $\sin x$ we see that

$$a_1 = -2 \cdot 3 \cdot a_3,$$
$$a_3 = -4 \cdot 5 \cdot a_5,$$

and so on.

But since $\sin x \to x$ as $x \to 0$ the first term of the series must be x, i.e. $a_1 = 1$ (§ 8).

$$\therefore \quad a_3 = -\frac{1}{3!}, \quad a_5 = \frac{1}{5!},$$

and so on.

Hence $$\sin x \equiv x - \frac{x^3}{3!} + \frac{x^5}{5!} - \dots,$$

and it is easily seen that the series is convergent for all values of x.

By substitution in the second equation or directly it can easily be shown that

$$\cos x \equiv 1 - \frac{x^2}{2!} + \frac{x^4}{4!} - \frac{x^6}{6!} + \dots.$$

§ 66. To find the series for $\sin^{-1} x$.

We know that the derivative of $\sin^{-1} x$ is

$$\frac{1}{\sqrt{1-x^2}}, \text{ i.e. } (1-x^2)^{-\frac{1}{2}}.$$

This can be expanded by the Binomial Theorem as long as x lies between ± 1, which is clearly the case if $\sin^{-1} x$ is to have any real meaning.

$$\therefore \quad (1-x^2)^{-\frac{1}{2}} \equiv 1 + \frac{1}{2} x^2 + \frac{1 \cdot 3}{2 \cdot 4} x^4 + \dots.$$

Since this series is the derivative of that for $\sin^{-1} x$ we must have

$$\sin^{-1} x = A + x + \frac{1}{2}\frac{x^3}{3} + \frac{1.3}{2.4}\frac{x^5}{5} + \frac{1.3.5}{2.4.6}\frac{x^7}{7} + \dots,$$

for this series when differentiated gives the series for $(1 - x^2)^{-\frac{1}{2}}$. The constant A is seen to be zero since $\sin^{-1} x = 0$ when $x = 0$.

Hence finally $\quad \sin^{-1} x \equiv x + \frac{1}{2}\frac{x^3}{3} + \frac{1.3}{2.4}\frac{x^5}{5} + \dots.$

In a precisely similar manner the series for $\tan^{-1} x$ may be found.

§ 67. Rolle's Theorem.

There is one general theorem, which, although it does not directly relate to the subject of Expansions, leads to many applications in this subject. This is known as Rolle's Theorem and may be stated as follows.

" *If $f(x)$ and $f'(x)$ are both continuous functions between $x = a$ and $x = b$ and if $f(a) = 0$, and $f(b) = 0$, then $f'(x)$ must be zero for at least one value of x lying between $x = a$ and $x = b$.*"

The theorem may be stated more conveniently, but perhaps less accurately in the form, "Between each of the roots of the equation $f(x) = 0$, there lies at least one root of the equation $f'(x) = 0$." In this form it is often given in treatises on the Theory of Equations. The theorem can be proved by reference to a figure, if we assume that the function can be represented by a graph. The theorem then states the obvious geometrical fact that for at least one point between any two points where the axis of x is crossed the tangent is parallel to the axis of x.

The various cases that may arise are illustrated by Fig. 21. Between A and B the theorem is not true owing to a discontinuity in the value of $f(x)$ occurring at P. Between B and C, also between C and D, the theorem is true in the ordinary way. Between D and E, and also between E and F, there are discontinuities in the value of $f'(x)$ which becomes infinite at S, and undergoes an abrupt change at T. Between F and G there is more than one point at which $f'(x) = 0$, but the reader will notice that in such a case there must be an odd number of such points. One type of breakdown is not included in

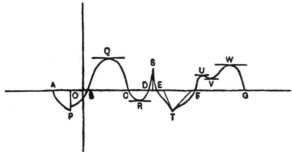

Fig. 21

the figure, viz. a point at which $f(x)$ becomes infinite, but this, as we have stated previously, counts as a discontinuity and is illustrated by what happens between A and B.

The student will also notice that the converse of the theorem is not necessarily true: e.g. between U and V, two points where $f'(x) = 0$, there is no point at which $f(x) = 0$ and there is no discontinuity either of $f(x)$ or $f'(x)$. [There is of course an inflexion, but this is not a discontinuity.]

An interesting application of Rolle's Theorem is to consider the function

$$f(x) - f(a) - (x - a)\frac{f(b) - f(a)}{b - a}.$$

This vanishes when $x = a$, and when $x = b$, and hence Rolle's Theorem may be applied. We thus obtain an independent proof of the First Theorem of the Mean given in § 70.

The reader should go through the work himself, as it is very important. It forms the first step of the proof of Taylor's Theorem (see Ex. VI, No. 32).

§ 68. Two General Theorems.

There are two general theorems, on the subject of the Expansion of functions in Infinite series, which we include here owing to their great utility. It can easily be seen that the extreme generality of these results precludes the possibility of a simple proof, which would be within the scope of this work. We therefore content ourselves with a statement of these theorems and of the conditions under which they may be assumed to hold. The student who is desirous of further information on the subject of these series, and points arising in connection with the proof of them, is referred to Goursat's *Cours d'Analyse*, or G. H. Hardy's *Course of Pure Mathematics*, in either of which a full discussion will be found.

§ 69. Taylor's Theorem.

Provided that all the derivatives employed exist and are continuous,

$$f(x + h) \equiv f(x) + hf'(x) + \frac{h^2}{2!} f''(x) + \dots \text{ ad inf.}$$

It is more correct to state the theorem in the "remainder" form, as follows :

$$f(x + h) = f(x) + hf'(x) + \frac{h^2}{2!} f''(x) + \dots + \frac{h^n}{n!} f^n(\xi),$$

where ξ is some value of x lying between x and $x + h$. If now it can be proved that this remainder $\dfrac{h^n}{n!} f^n(\xi) \to 0$ as $n \to \infty$ the first form of the series is valid, but although this is frequently true, the proof in any particular case is often a matter of some difficulty.

Example. $\sin(x + h)$.

$$f(x) = \sin x \text{ and hence } f^n(x) = (-1)^{n-1} \sin\left(x + \frac{n\pi}{2}\right) \quad (\S\,60),$$

since $\sin\left(x + \dfrac{n\pi}{2}\right) \not> 1$ and $\not< -1$, and $\dfrac{h^n}{n!} \to 0$ as $n \to \infty$,

the remainder $\dfrac{h^n}{n!} f^n(\xi) \to 0$ and the series is valid.

$$\therefore \ \sin(x + h) = \sin x + h \cos x - \frac{h^2}{2!} \sin x - \frac{h^3}{3!} \cos x + \ldots$$

$$= \sin x \left(1 - \frac{h^2}{2!} + \frac{h^4}{4!} - \ldots\right) + \cos x \left(h - \frac{h^3}{3!} + \frac{h^5}{5!} - \ldots\right)$$

$$= \sin x \cos h + \cos x \sin h.$$

Note that if $F(x)$ and $\phi(x)$ are both zero when $x = a$, then

$$\operatorname*{Lt}_{x \to a} F(x)/\phi(x) = \operatorname*{Lt}_{x \to a} F'(x)/\phi'(x).$$

§ 70. The Theorem of the Mean.

One particular case of this theorem is known as the "First Theorem of the Mean" and is easily seen to be true for any function that can be expressed by a continuous graph, and having a continuous derivative[*].

If we stop the series at the first derivative we get the form

$$f(x + h) = f(x) + hf'(\xi),$$

where ξ is some value between x and $x + h$.

---

[*] In point of fact the theorem is true whether $f'(x)$ is continuous or not; see Ex. VI, No. 32.

Writing this in the form

$$\frac{f(x+h)-f(x)}{h}=f'(\xi),$$

its geometrical significance is easily seen.

For if in Fig. 22 P is the point corresponding to the value x and P' to the value $(x+h)$, we see that

$$\frac{f(x+h)-f(x)}{h}=\frac{P'L}{PL}$$

= the slope of the chord PP'.

Also $f'(\xi)$ is the slope of the tangent at some intermediate point Q.

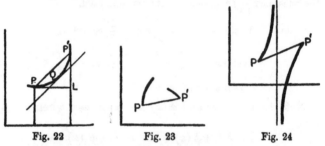

Fig. 22   Fig. 23   Fig. 24

The theorem thus expresses the geometrical fact that some point Q exists between P and P', such that the tangent at Q is parallel to the chord PP'.

It is easily seen from Figs. 23 and 24 that the theorem is not true if any discontinuity exists between P and P'.

The reader can easily construct figures for himself in which the theorem is true for *more* than one value of x.

§ 71. Maclaurin's Theorem.

If, in Taylor's Theorem, we substitute x for h, and 0 for x, we get a form of the result, which goes by the name of Maclaurin's Theorem.

This may be expressed in the form

$$f(x) = f(0) + xf'(0) + \frac{x^2}{2!}f''(0) + \ldots + \frac{x^n}{n!}f^n(0) + \ldots,$$

the notation $f^n(0)$ signifying that the nth derivative is to be calculated in terms of x, and then zero substituted for x. The validity of the expansion depends on a condition analogous to that of Taylor's Theorem, viz. that $\frac{x^n}{n!}f^n(\xi) \to 0$ as $n \to \infty$ where ξ here represents a value lying between 0 and x.

Example.

To expand $\log(1+x)$ by Maclaurin's Theorem:

$$f(x) = \log(1+x), \qquad\qquad \therefore \quad f(0) = 0,$$
$$f'(x) = (1+x)^{-1}, \qquad\qquad f'(0) = 1,$$
$$f''(x) = -1 \cdot (1+x)^{-2}, \qquad\qquad f''(0) = -1,$$
$$\ldots\ldots\ldots\ldots\ldots\ldots \qquad\qquad f'''(0) = 2!,$$
$$f^n(x) = (-1)^{n-1}(n-1)!(1+x)^{-n}, \quad f^n(0) = (-1)^{n-1}(n-1)!$$

Hence the expansion (if valid) is

$$f(x) = 0 + x \cdot 1 + \frac{x^2}{2!}(-1) + \frac{x^3}{3!} \cdot 2! + \ldots$$
$$+ \frac{x^n}{n!}(-1)^{n-1}(n-1)! + \ldots$$
$$= x - \frac{x^2}{2} + \frac{x^3}{3} - \ldots + (-1)^{n-1}\frac{x^n}{n} + \ldots.$$

We also see that the remainder is

$$\frac{x^n}{n!}(-1)^{n-1}(n-1)!(1+\xi)^{-n} = \frac{x^n}{(1+\xi)^n} \cdot \frac{(-1)^n}{n}.$$

Now ξ lies between 0 and x, and hence as long as x lies between $+1$ and -1 the remainder $\rightarrow 0$ as $n \rightarrow \infty$, but not otherwise.

Hence the series is valid if x lies between $+1$ and -1, as proved above.

§ 72. Maxima and Minima. Further Tests.

We are now in a position to complete our statement of the conditions governing maximum and minimum values.

For suppose $x = a$ is the value we desire to test and that applying the ordinary rules we find $f''(a) = 0$.

Now the essential condition for $x = a$ to give a maximum is that $f(a) > f(a \pm h)$ where h is a small quantity.

Let $x = a + h$ be a value close to $x = a$ [by making h positive or negative we can cause this to represent a value on either side of the critical value].

Now by Taylor's Theorem

$$f(a + h) = f(a) + hf'(a) + \frac{h^2}{2!}f''(a) + \frac{h^3}{3!}f'''(a) + \dots.$$

$$\therefore \quad f(a + h) - f(a) = \frac{h^3}{3!}f'''(a) + \frac{h^4}{4!}f^{iv}(a) + \dots,$$

for $\qquad\qquad f'(a) = 0, \ f''(a) = 0.$

Now if the value $x = a$ gives a true maximum or minimum value of y, $f(a + h) - f(a)$ will have the same sign whether h is positive or negative, and this sign will be negative for a maximum and positive for a minimum. But the term containing h^3 changes sign when h changes sign. Therefore $f'''(a)$ must vanish. Also for a maximum value $f^{iv}(a)$ must be negative and for a minimum it must be positive.

Following this argument we can see that the complete test for maxima and minima is as follows:

"*The first derivative which does not vanish must be one of even order; and for a maximum this derivative must be negative; for a minimum it must be positive."*

If the first non-vanishing derivative is of odd order the point considered is an inflexion on the graph of the function.

§ 73. Worked Example.

The following example illustrates a further method of finding expansions.

Find an expansion for $e^{\sin^{-1}x}$.

Let
$$y = e^{\sin^{-1}x},$$
$$y_1 = \frac{e^{\sin^{-1}x}}{\sqrt{1-x^2}}.$$
$$\therefore \quad (1-x^2)^{\frac{1}{2}}\, y_1 = y.$$

Differentiate again.

$$\therefore \quad (1-x^2)^{\frac{1}{2}}\, y_2 - \frac{x}{(1-x^2)^{\frac{1}{2}}}\, y_1 = y_1.$$
$$\therefore \quad (1-x^2)\, y_2 - x y_1 - y_1 (1-x^2)^{\frac{1}{2}} = y.$$
$$\therefore \quad (1-x^2)\, y_2 - x y_1 - y = 0.$$

Now if
$$y \equiv a_0 + a_1 x + a_2 x^2 + a_3 x^3 + \ldots + a_n x^n + \ldots,$$
then
$$y_1 \equiv a_1 + 2a_2 x + 3a_3 x^2 + \ldots + n a_n x^{n-1} + \ldots,$$
$$y_2 \equiv 2a_2 + 2\cdot 3a_3 x + \ldots + n(n-1) a_n x^{n-2} + \ldots.$$
$$\therefore \quad (1-x^2)\{2a_2 + 2\cdot 3a_3 x + \ldots + n(n-1) a_n x^{n-2} + \ldots\}$$
$$- x\{a_1 + 2a_2 x + 3a_3 x^2 + \ldots + n a_n x^{n-1} + \ldots\}$$
$$- \{a_0 + a_1 x + a_2 x^2 + \ldots + a_n x^n + \ldots\} \equiv 0.$$

Equating coefficients we have

$$2a_2 = a_0,$$
$$2\cdot 3a_3 = 2a_1,$$
$$3\cdot 4a_4 = 5a_2,$$
$$4\cdot 5a_5 - 2\cdot 3a_3 = 4a_2,$$

and in general

$$(n+2)(n+1)\, a_{n+2} - n(n-1)\, a_n - n a_n - a_n = 0,$$

i.e. $$(n+2)(n+1)\, a_{n+2} = (n^2+1)\, a_n.$$

By inspection $a_0 = 1,$

$$'a_1 = 1,$$

for we see from the series of § 66 that the first term must be x.

$$\therefore \quad a_2 = \frac{1}{2}, \qquad a_4 = \frac{5}{3 \cdot 4} \cdot \frac{1}{2}, \text{ etc.}$$

and $$a_3 = \frac{2}{2 \cdot 3}, \quad a_5 = \frac{2 \cdot 5}{2 \cdot 3 \cdot 4 \cdot 5}, \text{ etc.,}$$

and the expansion finally becomes

$$1 + x + \frac{x^2}{2!} + \frac{1(1+1^2)}{3!}\, x^3 + \frac{1(1+2^2)}{4!}\, x^4$$

$$+ \frac{1(1+1^2)(1+3^2)}{5!}\, x^5 + \dots \text{ ad inf.}$$

EXAMPLES VI

Obtain a form for the nth derivative of the following:

(1) x^{n+r}. (2) $\dfrac{1}{x^r}$. (3) $\log x$.

(4) $\dfrac{1}{(x+1)^2}$. (5) $\dfrac{1}{x^2+4x+3}$. (6) $\cos x$.

(7) $\sin^2 x$. (8) $e^x \cos x$. (9) $e^x \cos^2 x$.

(10) $\sin^3 x$ (use $\sin 3x$). (11) $\dfrac{1}{x^2-a^2}$.

(12) $\dfrac{x^2+1}{x^3-x}$.

Differentiate the following products n times by Leibnitz' Theorem:

(13) $x^2 e^x$. (14) $x^3 \cos x$. (15) $(1+x)^2 \log(1+x)$.

(16) $x^2 \dfrac{d^2y}{dx^2} - 2x \dfrac{dy}{dx} + (n^2+2)\, y$. (17) $e^x \sin x$.

(18) If $y = e^{\tan^{-1}x}$, prove that

$$(1+x^2)\frac{d^2y}{dx^2} + (2x-1)\frac{dy}{dx} = 0.$$

(19) If $y = (1+x^2)^{\frac{m}{2}}\sin(m\tan^{-1}x)$, prove that

$$(1+x^2)\frac{d^2y}{dx^2} - 2(m-1)x\frac{dy}{dx} + m(m-1)y = 0.$$

(20) Obtain the expansion of $e^{\tan^{-1}x}$ by using Ex. 18 and the method of § 73.

Obtain expansions of the following in terms of x, mentioning any restrictions placed on the variable:

(21) $\cos x$. (22) $\tan^{-1}x$. (23) $\log\dfrac{1+x}{1-x}$.

(24) $\sinh x = \dfrac{e^x - e^{-x}}{2}$. (25) $\log(x+\sqrt{1+x^2})$.

*(26) Prove that

$$e^x \sin x = x + x^2 + \frac{x^3}{3!}2^{\frac{3}{2}}\sin\frac{3\pi}{4} + \ldots + \frac{x^n}{n!}2^{\frac{n}{2}}\sin\frac{n\pi}{4} + \ldots.$$

*(27) If $y = \{\sqrt{1+x^2}+1\}^{\frac{1}{2}}$, prove that

$$(1+x^2)y_2 + xy_1 - \frac{y}{4} = 0.$$

Hence find the general term in the expansion of

$$\left(\frac{\sqrt{1+x^2}+1}{2}\right)^{\frac{1}{2}}.$$

*(28) If $y = (1+x)^m\log(1+x)$, prove that

$$(1+x)\frac{dy}{dx} - my = (1+x)^m.$$

Hence, assuming Maclaurin's Theorem, prove that the coefficient of x^{m+1} in the expansion of y in ascending powers of x is $\dfrac{1}{m+1}$.

*(29) It is desired to have series by which $\{\frac{1}{4}\pi - x\}\tan x$ can be calculated (1) when x is small, (2) when x is nearly equal to $\dfrac{\pi}{4}$. By Taylor's Theorem or otherwise, find the two series up to the terms involving cubes of small quantities.

(30) Prove that

$$\tfrac{1}{2}\log(1+x^2)\tan^{-1}x \equiv \frac{s_2x^3}{3} - \frac{s_4x^5}{5} + \frac{s_6x^7}{7} - \dots,$$

where

$$s_n \equiv \frac{1}{1} + \frac{1}{2} + \frac{1}{3} + \dots + \frac{1}{n}.$$

(31) Show that, when x is small, an approximate series for $x\cot x$ is

$$1 - \frac{x^2}{3} - \frac{x^4}{45}.$$

(32) By applying Rolle's Theorem to the function

$$f(a+h) - f(x) - \frac{a+h-x}{h}\left[f(a+h) - f(a)\right]$$

prove the first theorem of the mean analytically.

[Note. This proof does not require $f'(x)$ to be continuous but merely that it should exist.]

(33) By considering the first derivative of the function

$$f(a+h) - f(x) - (a+h-x)f'(x)$$
$$- \left(\frac{a+h-x}{h}\right)^2\{f(a+h) - f(a) - hf'(a)\},$$

prove the Second Theorem of the Mean.

*(**34**) Prove that if $f(x)$ is finite and continuous between $x=a$ and $x=b$, then

$$\frac{f(a)-f(b)}{a-b} = f'\left(\frac{a+b}{2} + \theta\frac{a-b}{2}\right),$$

where θ is a number lying between $+1$ and -1.

If $f(x)=x^3$, shew that as long as a and b have the same sign there is a unique value of θ; but in the case when $2a+b=0$, the equation is satisfied by $\theta=-\tfrac{1}{2}$ and 1. Illustrate this result graphically.

(35) If $x=a$ is an approximation to a root of an equation $f(x)=0$, show that $a - \dfrac{f(a)}{f'(a)}$ is sometimes a closer approximation.

Use this method to approximate to a root of $x^3 + 4x - 40 = 0$.

CHAPTER VII

MISCELLANEOUS APPLICATIONS TO ALGEBRA, TRIGONOMETRY AND MECHANICS

§ 74. Application to Algebraic Series.

We first give a method which does not seem to be very widely given in text-books on Algebra.

Example.

If c_0, c_1, c_2, ... c_n are coefficients of the series for $(1 + x)^n$, prove that

(i) $c_0 + 2c_1 + 3c_2 + ... + (n + 1)c_n \equiv 2^{n-1}(n + 2)$,

(ii) $\dfrac{c_0}{2} + \dfrac{c_1}{3} + \dfrac{c_2}{4} + ... + \dfrac{c_n}{n + 2} \equiv \dfrac{1 + n\,2^{n+1}}{(n + 1)(n + 2)}$.

(i) Since

$$(1 + x)^n \equiv c_0 + c_1 x + c_2 x^2 + ... + c_n x^n,$$

we have by differentiating both sides of this identity

$$n(1 + x)^{n-1} \equiv c_1 + 2c_2 x + 3c_3 x^2 + ... + nc_n x^{n-1}.$$

Hence

$$(1 + x)^n + n(1 + x)^{n-1}$$
$$\equiv c_0 + c_1 + (c_1 + 2c_2)x + ... + (c_{n-1} + nc_n)x^{n-1} + c_n x^n.$$

Put $x = 1$ and collect terms. We have

$$2^n + n\,2^{n-1} \equiv c_0 + 2c_1 + 3c_2 + ... + (n + 1)c_n.$$

But $\qquad\qquad 2^n + n\,2^{n-1} = 2^{n-1}(n + 2),$

and the identity is proved.

(ii) Let

$$f(x) \equiv \frac{c_0 x^2}{2} + \frac{c_1 x^3}{3} + \dots + \frac{c_n x^{n+2}}{n+2}.$$

$$\therefore \quad \frac{df(x)}{dx} \equiv c_0 x + c_1 x^2 + \dots + c_n x^{n+1}$$

$$\equiv x(1+x)^n$$

$$\equiv (1+x)^{n+1} - (1+x)^n.$$

Hence $f(x)$ is a function such that its derivative is

$$(1+x)^{n+1} - (1+x)^n.$$

It must therefore be

$$\frac{(1+x)^{n+2}}{n+2} - \frac{(1+x)^{n+1}}{n+1} + C,$$

where C is some constant (for we have seen that a constant has no derivative).

Since $f(x) = 0$ when $x = 0$, we see that the value of C is

$$\frac{1}{n+1} - \frac{1}{n+2} = \frac{1}{(n+2)(n+1)}.$$

Hence finally,

$$\frac{c_0 x^2}{2} + \frac{c_1 x^3}{3} + \dots + \frac{c_n x^{n+2}}{n+2}$$

$$\equiv \frac{(1+x)^{n+2}}{n+2} - \frac{(1+x)^{n+1}}{n+1} + \frac{1}{(n+2)(n+1)}.$$

Now put $x = 1$.
Hence

$$\frac{c_0}{2} + \frac{c_1}{3} + \dots + \frac{c_n}{n+2} \equiv \frac{2^{n+2}}{n+2} - \frac{2^{n+1}}{n+1} + \frac{1}{(n+2)(n+1)}$$

$$\equiv \frac{2^{n+1}(2n+2-n-2)+1}{(n+2)(n+1)}$$

$$\equiv \frac{n2^{n+1}+1}{(n+2)(n+1)}.$$

In this way, and also by using other values for x, many identities can be found. The student who is acquainted with the Integral Calculus will recognize the last example as one of integration*.

§ 75. Trigonometrical Series.

A similar method can often be applied to trigonometrical series, especially in deriving other series from a known series.

Thus from

$$\sin \theta \equiv \theta \left\{ 1 - \frac{\theta^2}{\pi^2} \right\} \left\{ 1 - \frac{\theta^2}{2^2 \pi^2} \right\} \dots ad \ inf.$$

we derive by logarithmic differentiation the series for $\cot \theta$, as follows:

$$\log \sin \theta \equiv \log \theta + \log \left(1 - \frac{\theta^2}{\pi^2} \right) + \log \left(1 - \frac{\theta^2}{2^2 \pi^2} \right) + \dots.$$

Hence, differentiating both sides, we have

$$\cot \theta \equiv \frac{1}{\theta} - \frac{2\theta}{\pi^2 - \theta^2} - \frac{2\theta}{2^2 \pi^2 - \theta^2} - \dots ad \ inf.$$

§ 76. Applications to Mechanics.

There are many important applications of the Differential Calculus to Mechanics. We have already pointed out in Chapter II that $\frac{ds}{dt}$ represents velocity, and $\frac{d^2 s}{dt^2}$ acceleration. We give in this and the following sections a few more difficult applications. The scope of this work obviously prevents these from being given in very great detail, and for further information the student is recommended to study the standard works on Mechanics.

* The student is advised to try by this method Ex. **xxv** in C. Smith's *Treatise on Algebra.*

§ 77. Stability of Equilibrium.

Questions concerning stability of equilibrium are often best solved by obtaining the condition that the height of the centre of gravity* of the body should be maximum or minimum. Clearly if the configuration of the body we are studying is such as to make this height a minimum any small displacement will tend to make the body return to that position, and hence the equilibrium will be stable; and in the same way a maximum value will denote unstable equilibrium. In investigating this we consider the height of the centre of gravity when the body is slightly displaced from its equilibrium position; otherwise there is no variable in the expression and it cannot be differentiated. The following example should illustrate the method sufficiently.

§ 78. Worked Example.

A cube rests on the curved surface of a solid hemisphere, which is rough enough to prevent sliding. Find the dimensions of the largest cube for which the equilibrium is stable.

Consider the system in a slightly displaced position such that the normal to the sphere at its point of contact with the cube makes a small angle θ with the vertical.

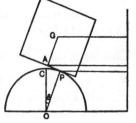

Fig. 25

Let the radius of the hemisphere be r and a side of the cube a.

Since the cube does not slide, the point of contact will move equal distances along the side of the cube and along the arc.

* In more difficult examples use the Potential Energy of the System.

Hence $PA = PC = r\theta$.

Using the method of projections, the height of the centre of gravity of the cube is equal to the sum of the projections of OP, AP, and AG on a vertical line.

$$\therefore \quad h = r \cos \theta + r\theta \sin \theta + \frac{a}{2} \cos \theta.$$

For equilibrium h is maximum or minimum.

$$\therefore \quad \frac{dh}{d\theta} = -r \sin \theta + r\theta \cos \theta + r \sin \theta - \frac{a}{2} \sin \theta = 0.$$

Therefore equilibrium occurs when

$$r\theta \cos \theta = \frac{a}{2} \sin \theta.$$

The only obvious solution is $\theta = 0$.

Also

$$\frac{d^2h}{d\theta^2} = r \cos \theta - r\theta \sin \theta - \frac{a}{2} \cos \theta$$

$$= \left(r - \frac{a}{2} \right) \cos \theta - r\theta \sin \theta.$$

Thus, when $\theta = 0$,

$$\frac{d^2h}{d\theta^2} = r - \frac{a}{2}.$$

Hence, so long as $r > \frac{a}{2}$, $\frac{d^2h}{d\theta^2}$ is positive and the height of the c.g. is a minimum; that is, equilibrium is stable: and when $r < \frac{a}{2}$, the same argument proves the equilibrium unstable.

To investigate the case when $r = \frac{a}{2}$, we have to apply the test given in § 72.

In this case $\dfrac{d^2h}{d\theta^2} = -r\theta\sin\theta,$

$$\therefore \quad \dfrac{d^3h}{d\theta^3} = -r\sin\theta - r\theta\cos\theta,$$

which also vanishes when $\theta = 0$,

and $\dfrac{d^4h}{d\theta^4} = -r\cos\theta + r\theta\sin\theta - r\cos\theta$

$$= r\theta\sin\theta - 2r\cos\theta,$$

which is *negative* when $\theta = 0$.

Hence when $r = \dfrac{a}{2}$, the equilibrium is *unstable*.

Therefore the largest cube is one whose side is just less than the diameter of the sphere.

§ 79. Virtual Work.

The application of this principle to Statics is known as the principle of Virtual Work. Since positions of equilibrium occur when the Potential Energy is at a maximum or minimum it follows that, if we give a system a small displacement from its equilibrium position, the change in the Potential Energy will be a small quantity of the second order. Thus we may equate to zero the first-order elements which occur in the expression for the increase of the potential energy. This gives rise to an equation which may enable us to find an unknown force or constraint of the system.

It should further be noted that we can leave out all forces such as reactions at hinges, points of contact with smooth bodies, etc.; these forces being proved, in the general theory, to contribute nothing to the result. Strings, in examples, have usually to be considered as being slightly elastic. The displacements required may be obtained by differentiating.

The principle, and the method of applying it, are illustrated in the following article.

§ 80. Worked Example.

A square framework formed of equal uniform rods, each of weight W, is hung up freely by one corner. A weight W is suspended from each of the lower corners, and the square shape is preserved by a light rod along the horizontal diagonal.

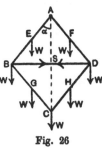

Find the stress in this rod.

Consider the system, not in the equilibrium position, but in a position where all the rods are inclined to the vertical at an angle α. It is seen that in the equilibrium position $\alpha = 45°$.

Fig. 26

Let the length of each rod be denoted by $2a$. Then the depths below the level of A are:

$$a \cos \alpha \quad \text{for} \quad E \text{ and } F,$$
$$2a \cos \alpha \quad \text{for} \quad B \text{ and } D,$$
$$3a \cos \alpha \quad \text{for} \quad G \text{ and } H,$$

and

$$4a \cos \alpha \quad \text{for} \quad C,$$

and the length of the supporting rod BD is $4a \sin \alpha$.

Next let the angle α increase by a small increment $\delta\alpha$. Then, differentiating, we see that the corresponding changes in these depths are respectively, to the first order:

$$\text{for } E \text{ and } F \quad - \ a \sin \alpha \,.\, \delta\alpha,$$
$$\text{for } B \text{ and } D \quad - 2a \sin \alpha \,.\, \delta\alpha,$$
$$\text{for } G \text{ and } H \quad - 3a \sin \alpha \,.\, \delta\alpha,$$
$$\text{for } C \qquad\quad - 4a \sin \alpha \,.\, \delta\alpha,$$

while the corresponding increment in BD is

$$4a \cos \alpha . \delta\alpha.$$

Notice that we need not consider the horizontal displacements of E, F, B, D, G, H and C, as these are not in the directions of the forces at these points, and hence contribute nothing to the work done, and similarly for the vertical displacement of the rod BD.

Hence the total work done in making this displacement is

$$- 2W a \sin \alpha \, \delta\alpha - 4W a \sin \alpha \, \delta\alpha - 6W a \sin \alpha \, \delta\alpha$$
$$- 4W a \sin \alpha \, \delta\alpha + 4S a \cos \alpha \, \delta\alpha.$$

Since the work done is a small quantity of the second order, the coefficient of $\delta\alpha$ vanishes.

$$\therefore \quad 4S a \cos \alpha = 16 W a \sin \alpha.$$
$$\therefore \quad S = 4W \tan \alpha.$$

But in the position of equilibrium $\alpha = 45°$.

$$\therefore \quad S = 4W.$$

§ 81. Connected Displacements.

It often happens that the positions of the points cannot all be determined in terms of one angle, and two or more have to be introduced. These angles, or parameters, will be found to be connected by a relation, or relations, derived from the geometry of the figure. This relation (or relations) will have to be written down and differentiated, before we can successfully complete the example. This is illustrated by the following example.

Consider the case of a pentagon of rigid rods supported by a tie-bar as in the figure and suspended from one vertex. It is required to find the stress in this bar.

Let the inclinations of AB and BC to the vertical be respectively α and β, and the side of the pentagon be $2a$. The equilibrium position is given when $\alpha = 54°$, $\beta = 18°$. The depth below A of F and L is $a \cos \alpha$, that of G and K, $2a \cos \alpha + a \cos \beta$, that of H, $2a \cos \alpha + 2a \cos \beta$, whilst the length of the rod BE is $4a \sin \alpha$.

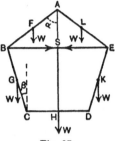

Fig. 27

If the system is slightly deformed in a symmetrical manner so that CD remains horizontal and H exactly below A, then we can obtain the displacements of the points by differentiation as before.

Therefore, calculating the depths below A, we see that the displacement of F and L is $-a \sin \alpha \, \delta\alpha$,

,, G and K is $-2a \sin \alpha \, \delta\alpha - a \sin \beta \, \delta\beta$,

,, H is $-2a \sin \alpha \, \delta\alpha - 2a \sin \beta \, \delta\beta$,

while the alteration in the length of BE is $4a \cos \alpha \, \delta\alpha$.

Hence, equating to zero the work done, we have

$$W \{8a \sin \alpha \, \delta\alpha + 4a \sin \beta \, \delta\beta\} = S \{4a \cos \alpha \, \delta\alpha\}.$$

But from geometrical considerations

$$2a \sin \alpha - 2a \sin \beta = CH = a.$$

$$\therefore \quad \cos \alpha \, \delta\alpha - \cos \beta \, \delta\beta = 0,$$

by differentiating this.

Hence $\delta\alpha \propto \cos \beta$,

and $\delta\beta \propto \cos \alpha$.

Substituting in the equation of virtual work, we get

$$W \{8 \sin \alpha \cos \beta + 4 \sin \beta \cos \alpha\} = S \{4 \cos \alpha \cos \beta\}.$$

$$\therefore \quad S = W\frac{2\sin\alpha\cos\beta + \sin\beta\cos\alpha}{\cos\alpha\cos\beta}$$

$$= W\frac{\sin(\alpha+\beta) + \sin\alpha\cos\beta}{\cos\alpha\cos\beta}$$

$$= W\frac{\sin 72° + \sin 54°\cos 18°}{\cos 54°\cos 18°}$$

$$= W\frac{1 + \sin 54°}{\cos 54°},$$

which reduces to $W\cot 18°$.

§ 82. Motion on a curve.

Another case in which the Calculus is employed is in discovering the conditions of motion on a curve. This principle may be employed to find the ordinary conditions of motion in a circle. It will be seen however that, up to a certain point, the method is perfectly general, and applicable to any other curve.

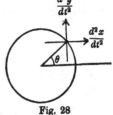

Fig. 28

Now we know from Analytical Geometry that the coordinates of any point on the circle may be expressed parametrically in the form

$$x = a\cos\theta, \quad y = a\sin\theta.$$

Differentiating once with respect to t, we get

$$\left.\begin{aligned} \frac{dx}{dt} &= -a\sin\theta\,\frac{d\theta}{dt}\\ \frac{dy}{dt} &= a\cos\theta\,\frac{d\theta}{dt} \end{aligned}\right\}.$$

These are the components of the velocity resolved along lines parallel to the axes of coordinates.

Compounding these, we see that the magnitude of the

velocity is $a\dfrac{d\theta}{dt}$, and also that its direction makes an angle $\tan^{-1}(-\cot\theta)$ with the axis of x. But this is seen to be perpendicular to the radius, and hence the velocity is directed along the tangent at the point.

To find the components of acceleration we have to differentiate again, getting

$$\left.\begin{aligned}
\frac{d^2x}{dt^2} &= -a\sin\theta\frac{d^2\theta}{dt^2} - a\cos\theta\left(\frac{d\theta}{dt}\right)^2\\
\frac{d^2y}{dt^2} &= a\cos\theta\frac{d^2\theta}{dt^2} - a\sin\theta\left(\frac{d\theta}{dt}\right)^2
\end{aligned}\right\}.$$

Hence the resolved part of the acceleration along the normal is equal to

$$\frac{d^2x}{dt^2}\cos\theta + \frac{d^2y}{dt^2}\sin\theta$$

$$= -a\left(\frac{d\theta}{dt}\right)^2\{\cos^2\theta + \sin^2\theta\}$$

$$= -a\left(\frac{d\theta}{dt}\right)^2$$

$$= -\frac{V^2}{a}.$$

Hence the normal component of acceleration is $-\dfrac{V^2}{a}$ **and this is true whether the velocity with which the circle is described is constant or not.**

There is also seen to be a component of the acceleration along the tangent, its value being

$$-\frac{d^2x}{dt^2}\sin\theta + \frac{d^2y}{dt^2}\cos\theta,$$

which reduces to

$$a\frac{d^2\theta}{dt^2}.$$

This vanishes when $\dfrac{d^2\theta}{dt^2}=0$, **i.e. when** $\dfrac{d\theta}{dt}$ **is constant,**

that is, if the circle is described with constant velocity.

Considerations of a precisely similar nature will enable us to investigate motion on any kind of curve.

EXAMPLES VII

Given that
$$(1+x)^n = c_0 + c_1 x + c_2 x^2 + \ldots + c_n x^n,$$
prove the following identities:

(1) $c_1 + 2c_2 + 3c_3 + \ldots + n c_n = n 2^{n-1}.$

(2) $c_0 - 2c_1 + 3c_2 - 4c_3 + \ldots + (-1)^n (n+1) c_n = 0.$

(3) $c_0 + \dfrac{c_1}{2} + \dfrac{c_2}{3} + \ldots + \dfrac{c_n}{n+1} = \dfrac{2^{n+1}-1}{n+1}.$

(4) $c_0 - \dfrac{c_1}{2} + \dfrac{c_2}{3} - \ldots + (-1)^n \dfrac{c_n}{n+1} = \dfrac{1}{n+1}.$

(5) Given the identity
$$\cos\frac{\theta}{2}\cos\frac{\theta}{2^2}\cos\frac{\theta}{2^3}\ldots\cos\frac{\theta}{2^n} = \frac{\sin\theta}{2^n \sin\dfrac{\theta}{2^n}},$$
deduce

(a) $\dfrac{1}{2}\tan\dfrac{\theta}{2} + \dfrac{1}{2^2}\tan\dfrac{\theta}{2^2} + \ldots + \dfrac{1}{2^n}\tan\dfrac{\theta}{2^n} = \dfrac{1}{2^n}\cot\dfrac{\theta}{2^n} - \cot\theta,$

(b) $\dfrac{1}{2^2}\sec^2\dfrac{\theta}{2} + \dfrac{1}{2^4}\sec^2\dfrac{\theta}{2^2} + \ldots + \dfrac{1}{2^{2n}}\sec^2\dfrac{\theta}{2^n}$
$$= \operatorname{cosec}^2\theta - \dfrac{1}{2^{2n}}\operatorname{cosec}^2\dfrac{\theta}{2^n}.$$

(6) A hemisphere rests on a fixed sphere of equal radius. Show that the equilibrium is unstable if the curved surfaces are in contact, but stable if the flat surface of the hemisphere is in contact with the sphere.

(7) Discuss the stability of a right circular cone resting on the curved surface of a hemisphere (the base of the cone touching the hemisphere).

(8) An isosceles triangle is placed in a vertical plane with one of its equal sides in contact with a sphere. If a is the length of the equal sides and α the vertical angle, prove that the equilibrium is stable if $\sin \alpha < \dfrac{3r}{a}$.

Discuss the case when $\sin \alpha = \dfrac{3r}{a}$.

(9) A square rests in a vertical plane between two pegs at the same horizontal level, and at a distance apart equal to half the diagonal of the square. Prove that there are three possible positions of equilibrium, but that only one is stable.

(10) A regular hexagon is formed of equal heavy rods whose weight is W.

It is suspended with one side horizontal and the shape is maintained by two light rods of suitable length connecting the extremities of the horizontal side with the vertices vertically below these extremities. Find the tension in these rods.

(11) The hexagon of the last example is hung up by one vertex and maintained in equilibrium by a rod joining the mid-point of the two vertical sides, which are not allowed to turn. Find the stress in the tie-rod.

(12) A square of uniform heavy rods is hung up by one corner and equilibrium maintained by a rod connecting the mid-points of the two lower rods. Find the force in this rod.

(13) A particle describes the cycloid
$$x = a\,(\theta + \sin \theta),$$
$$y = a\,(1 - \cos \theta)$$
in such a way that θ increases uniformly. Prove that the velocity at any point is proportional to $\cos \dfrac{\theta}{2}$, and that the acceleration is constant, but alters its direction uniformly.

(14) A particle describes the curve $ay^2 = x^3$. Show that there are three points at which the velocity is directed towards, or away from, the point $(a, 0)$.

If the velocity along the axis of y is proportional to x, show that the acceleration along the axis of x is constant.

(15) A particle describes the curve $y = \sin x$, the velocity in the direction of the y axis being a constant u.

Show that the acceleration along the other axial direction is

$$\frac{u^2 \sin x}{\cos^3 x}.$$

(16) A point describes the curve

$$x^4 + y^4 - 6x^2y^2 = 1,$$

in such a way that the acceleration is always towards the origin.

If r be the distance of a point from O, show that the velocity at any point $\propto r^3$ and that the acceleration is $\dfrac{3v^2}{r}$.

(17) The pentagon (§ 81) stands on its horizontal side, instead of being suspended. Find the tension in the connecting rod in this case.

(18) A hexagon consisting of six equal heavy rods is fixed in a position where one rod is horizontal and the regular shape is maintained by a string connecting the mid-point of this rod with the mid-point of the opposite rod. Find the tension in this string.

*(19) A particle moves along a straight line in such a way that at the end of t secs. its distance from a fixed point of the line is given by $s = 7t + 2 \cos 3t$.

Prove that the velocity of the particle never changes its direction, but that the acceleration vanishes whenever the velocity is 13 or 1 units.

*(20) A particle moves in a straight line so that its distance x from a fixed point of the line at time t is given by

$$x = Ae^{-kt} \cos nt.$$

Find expressions for its velocity v and acceleration f, at time t, and prove that they satisfy the equation

$$f + 2kv + (k^2 + n^2) x = 0.$$

(21) A point moves in a straight line in such a way that when it has travelled a distance s its velocity varies inversely as $(s + a)$, where a is a constant.

Prove that its retardation varies as the cube of the velocity, and show that the kinetic energy is diminishing at a rate proportional to the fourth power of the velocity.

CHAPTER VIII

CURVATURE OF PLANE CURVES

§ 83. Curvature of a Curve.

In many cases it is desirable to have some method of measuring the curvature, or rate of bending, of a given curve. A method which naturally suggests itself is the following.

Let P and P' be two points on a curve close to one another and let the two tangents at these points be drawn. The rate of bending of the curve can now be measured by comparing the change in the direction of the tangents with the length of the arc intercepted between the two points of contact. The greater the change in the direction of these

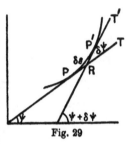

Fig. 29

tangents, the greater the curvature and vice versa. Adopting the usual mode of notation, we naturally call the inclinations of the tangents ψ and $\psi + \delta\psi$, and the small arc PP' δs.

The fraction $\dfrac{\delta\psi}{\delta s}$ thus gives an expression for the average rate of bending in passing from P to P'; by making the interval between P and P' smaller, and taking the limit, we see that an appropriate measure of the rate of bending, or curvature, *at* P is $\dfrac{d\psi}{ds}$.

In general this is a quantity which varies from point to

point of the curve, but there is one curve for which it is constant, viz. the circle.

§ 84. Curvature of the Circle. "Radius of Curvature."

It is easily seen that the curvature at all points of a circle is the same, and is equal to the reciprocal of the radius.

For the angle RTQ between two tangents is equal to the angle POQ at the centre.

$$\therefore \quad \frac{\angle RTQ}{\text{arc } PQ} = \frac{\theta}{r\theta} = \frac{1}{r}.$$

Fig. 30

By analogy with the circle we usually calculate not $\frac{d\psi}{ds}$ but its reciprocal $\frac{ds}{d\psi}$. This is known as "the radius of curvature" of the curve at the point under consideration and is always denoted by the Greek letter ρ.

§ 85. Standard Forms for ρ.

In most curves s and ψ are not easily found, and we have first to transform the expression $\frac{ds}{d\psi}$ into forms in which it is easy to deal with Cartesian, polar, or parametric coordinates. We give these transformations in full. The student should work through them carefully for himself and then memorize the results as formulae.

(a) Cartesian Coordinates.

We have seen that $\frac{dx}{ds} = \cos \psi$.

$$\therefore \quad \frac{ds}{dx} = \sec \psi.$$

$$\therefore \quad \frac{ds}{d\psi} = \frac{ds}{dx}\frac{dx}{d\psi} = \sec \psi \frac{dx}{d\psi}.$$

Again $\qquad \tan \psi = \dfrac{dy}{dx}.$

Therefore, differentiating with respect to x, we get

$$\sec^2 \psi \frac{d\psi}{dx} = \frac{d^2 y}{dx^2}.$$

$$\therefore \quad \frac{dx}{d\psi} = \frac{\sec^2 \psi}{\dfrac{d^2 y}{dx^2}}.$$

Substituting for $\dfrac{dx}{d\psi}$, we get

$$\frac{ds}{d\psi} = \frac{\sec^3 \psi}{\dfrac{d^2 y}{dx^2}} = \frac{(1 + \tan^2 \psi)^{\frac{3}{2}}}{\dfrac{d^2 y}{dx^2}}.$$

$$\therefore \quad \rho = \pm \frac{\left\{ 1 + \left(\dfrac{dy}{dx} \right)^2 \right\}^{\frac{3}{2}}}{\dfrac{d^2 y}{dx^2}} \qquad \ldots\ldots\text{(B)}.$$

Note. The ambiguous sign has been placed before the expression for ρ, as in many text-books. The usual convention of sign is to regard a curve such as that of Fig. 29, which presents its convex side downwards, as being of positive curvature, and vice versa. In fact if ψ is increasing with the usual convention, ρ is taken to be positive and vice versa. In the remaining forms the double sign will be omitted, or rather understood in the fractional index.

(b) Parametric Coordinates.

If x and y are both given as functions of a parameter t, we have seen that

$$\frac{dy}{dx} = \frac{\dfrac{dy}{dt}}{\dfrac{dx}{dt}}.$$

Then
$$\frac{d^2y}{dx^2} = \frac{d}{dx}\left\{\frac{dy}{dx}\right\}$$

$$= \left\{\frac{d}{dt}\frac{dy}{dx}\right\}\frac{dt}{dx}$$

$$= \frac{\dfrac{d}{dt}\left\{\dfrac{\dfrac{dy}{dt}}{\dfrac{dx}{dt}}\right\}}{\dfrac{dx}{dt}}$$

$$= \frac{\dfrac{d^2y}{dt^2}\dfrac{dx}{dt} - \dfrac{d^2x}{dt^2}\dfrac{dy}{dt}}{\left(\dfrac{dx}{dt}\right)^3} \qquad (\S\ 28,\ E).$$

Therefore substituting in formula (B),

$$\rho = \frac{\left\{1 + \dfrac{\left(\dfrac{dy}{dt}\right)^2}{\left(\dfrac{dx}{dt}\right)^2}\right\}^{\frac{3}{2}}\left(\dfrac{dx}{dt}\right)^3}{\dfrac{d^2y}{dt^2}\dfrac{dx}{dt} - \dfrac{d^2x}{dt^2}\dfrac{dy}{dt}}$$

$$= \frac{\left\{\left(\dfrac{dy}{dt}\right)^2 + \left(\dfrac{dx}{dt}\right)^2\right\}^{\frac{3}{2}}}{\dfrac{d^2y}{dt^2}\dfrac{dx}{dt} - \dfrac{d^2x}{dt^2}\dfrac{dy}{dt}} \qquad \ldots\ldots(C).$$

The attention of the student is called to the method of finding $\frac{d^2y}{dx^2}$.

(c) **Polar Coordinates.**

Here we know that $\psi = \phi + \theta$ (§ 57).

Hence
$$\frac{1}{\rho} = \frac{d\psi}{ds} = \frac{d\phi}{ds} + \frac{d\theta}{ds}.$$

Now
$$r\frac{d\theta}{ds} = \sin\phi.$$

$$\therefore \; \frac{1}{\rho} = \frac{\sin\phi}{r} + \frac{d\phi}{ds}$$

$$= \frac{\sin\phi}{r} + \frac{d\phi}{d\theta}\frac{d\theta}{ds}$$

$$= \frac{\sin\phi}{r}\left\{1 + \frac{d\phi}{d\theta}\right\}.$$

But
$$\phi = \tan^{-1} r\frac{d\theta}{dr}$$

$$= \tan^{-1}\frac{r}{\dfrac{dr}{d\theta}}.$$

$$\therefore \; \frac{d\phi}{d\theta} = \frac{1}{1 + \dfrac{r^2}{\left(\dfrac{dr}{d\theta}\right)^2}}\; \frac{\left(\dfrac{dr}{d\theta}\right)^2 - r\dfrac{d^2r}{d\theta^2}}{\left(\dfrac{dr}{d\theta}\right)^2}$$

$$= \frac{\left(\dfrac{dr}{d\theta}\right)^2 - r\dfrac{d^2r}{d\theta^2}}{r^2 + \left(\dfrac{dr}{d\theta}\right)^2},$$

and also
$$\sin\phi = \frac{r}{\sqrt{r^2 + \left(\dfrac{dr}{d\theta}\right)^2}},$$

since
$$\tan\phi = \frac{r}{\dfrac{dr}{d\theta}}.$$

Hence, substituting for $\sin \phi$ and $\dfrac{d\phi}{d\theta}$, we get

$$\frac{1}{\rho} = \frac{1}{\sqrt{r^2 + \left(\dfrac{dr}{d\theta}\right)^2}} \left\{ \frac{r^2 + 2\left(\dfrac{dr}{d\theta}\right)^2 - r\dfrac{d^2r}{d\theta^2}}{r^2 + \left(\dfrac{dr}{d\theta}\right)^2} \right\}$$

$$= \frac{r^2 + 2\left(\dfrac{dr}{d\theta}\right)^2 - r\dfrac{d^2r}{d\theta^2}}{\left\{ r^2 + \left(\dfrac{dr}{d\theta}\right)^2 \right\}^{\frac{3}{2}}}.$$

Hence $\qquad \rho = \dfrac{\left\{ r^2 + \left(\dfrac{dr}{d\theta}\right)^2 \right\}^{\frac{3}{2}}}{r^2 + 2\left(\dfrac{dr}{d\theta}\right)^2 - r\dfrac{d^2r}{d\theta^2}}$ $\qquad$ ......(D).

§ 86. Secondary Formulae for ρ.

Owing to the difficulty of remembering formula (D) of the last article, another form is often used which can be obtained as follows.

(d) "Pedal" Form.

As before $\psi = \phi + \theta$.

$$\therefore \quad \frac{1}{\rho} = \frac{d\psi}{ds} = \frac{d\psi}{dr}\frac{dr}{ds}$$

$$= \cos \phi \left\{ \frac{d\psi}{dr} \right\}$$

$$= \cos \phi \left\{ \frac{d\theta}{dr} + \frac{d\phi}{dr} \right\}.$$

But $\qquad r\dfrac{d\theta}{dr} = \tan \phi.$

$$\therefore \quad \frac{1}{\rho} = \cos \phi \left\{ \frac{\tan \phi}{r} + \frac{d\phi}{dr} \right\}$$

$$= \frac{\sin \phi}{r} + \cos \phi \frac{d\phi}{dr}.$$

Now $p = r \sin \phi$.

$$\frac{dp}{dr} = r \cos \phi \frac{d\phi}{dr} + \sin \phi.$$

Hence

$$\frac{1}{\rho} = \frac{1}{r}\frac{dp}{dr},$$

or

$$\rho = r\frac{dr}{dp} \qquad \ldots\ldots(E).$$

Now in many curves the "Pedal" (p, r) equation is fairly simple and this formula often supplies the best method of finding ρ in a Polar curve.

One further formula is given for ρ.

(e) "Tangential-Polar" Form.

We have seen in a previous chapter (§ 53) that p and $\frac{dp}{d\psi}$ are the perpendiculars from the origin on the tangent and normal respectively, and it is clear from the figure that

$$r^2 = p^2 + \left(\frac{dp}{d\psi}\right)^2.$$

Differentiate both sides of this equation with respect to p.

Then

$$2r\frac{dr}{dp} = 2p + 2\frac{dp}{d\psi}\frac{d}{dp}\left\{\frac{dp}{d\psi}\right\}.$$

$$\therefore \quad r\frac{dr}{dp} = \rho = p + \frac{dp}{d\psi}\frac{d}{dp}\left\{\frac{dp}{d\psi}\right\}$$

$$= p + \frac{d}{d\psi}\left\{\frac{dp}{d\psi}\right\}$$

$$= p + \frac{d^2p}{d\psi^2} \qquad \ldots\ldots(F).$$

The equation connecting p and ψ is sometimes called the "Tangential-Polar" Equation of a curve.

§ 87. Table of Formulae for ρ.

We collect here in a table the formulae for ρ we have just proved.

$$\rho = \text{(A)} \quad \frac{ds}{d\psi},$$

$$\text{(B)} \quad \frac{\left\{1 + \left(\dfrac{dy}{dx}\right)^2\right\}^{\frac{3}{2}}}{\dfrac{d^2y}{dx^2}},$$

$$\text{(C)} \quad \frac{\left\{\left(\dfrac{dy}{dt}\right)^2 + \left(\dfrac{dx}{dt}\right)^2\right\}^{\frac{3}{2}}}{\dfrac{d^2y}{dt^2}\dfrac{dx}{dt} - \dfrac{d^2x}{dt^2}\dfrac{dy}{dt}},$$

$$\text{(D)} \quad \frac{\left\{r^2 + \left(\dfrac{dr}{d\theta}\right)^2\right\}^{3}}{r^2 + 2\left(\dfrac{dr}{d\theta}\right)^2 - r\dfrac{d^2r}{d\theta^2}},$$

$$\text{(E)} \quad r\frac{dr}{dp},$$

$$\text{(F)} \quad p + \frac{d^2p}{d\psi^2}.$$

We see from formula (B) that, at a point of inflexion, (§ 45) at which we know that $\dfrac{d^2y}{dx^2} = 0$, ρ becomes infinite. In the same way, in formulae (C) and (D) the vanishing of the denominator gives us conditions for a point of inflexion in other coordinates.

§ 88. Worked Example. Curvature at Origin.

The following example illustrates an interesting method of obtaining the radius of curvature of an algebraic curve at the origin.

Find the radius of curvature at O of the conic

$$y + 2x = x^2 + 4xy + y^2.$$

Since y and x are connected, there must be an expansion of y in terms of x by Maclaurin's Theorem, viz.:

$$y = y_0 + x\left(\frac{dy}{dx}\right)_0 + \frac{x^2}{2!}\left(\frac{d^2y}{dx^2}\right)_0 + \dots.$$

Denote $\left(\frac{dy}{dx}\right)_0$ by p, and $\left(\frac{d^2y}{dx^2}\right)_0$ by q, and of course $y_0 = 0$.

Substitute $px + \frac{qx^2}{2!} + \dots$ for y in the equation to the curve.

This should give an identity, and hence we may equate coefficients of successive powers of x.

The equation becomes

$$px + \frac{qx^2}{2!} + \dots + 2x$$

$$\equiv x^2 + 4x\left\{px + \frac{qx^2}{2!} + \dots\right\} + \left\{px + \frac{qx^2}{2!} + \dots\right\}^2.$$

Equating the first two coefficients we have

$$\left.\begin{array}{l} p + 2 = 0 \\ \frac{q}{2} = 1 + 4p + p^2 \end{array}\right\},$$

whence

$$p = -2, \quad q = -6.$$

$$\rho = \frac{\left\{1 + \left(\frac{dy}{dx}\right)^2\right\}^{\frac{3}{2}}}{\frac{d^2y}{dx^2}} = \frac{(1 + p^2)^{\frac{3}{2}}}{q}$$

$$= -\frac{5^{\frac{3}{2}}}{6}$$

$$= -\frac{5\sqrt{5}}{6}.$$

Since we can easily transform to any point as origin, this method will evaluate ρ for any known point on an algebraic curve, but the work may easily become laborious.

EXAMPLES VIII A

Find ρ for any point of the following curves:

(1) $s = 4a \sin \psi$.

(2) $e^{\frac{s}{a}} = \sin \psi$.

(3) $y = c \cosh \dfrac{x}{c}$.

(4) $y = c \log \sec \dfrac{x}{c}$.

(5) $\left.\begin{array}{l} x = a\,(\theta + \sin \theta) \\ y = a\,(1 - \cos \theta) \end{array}\right\}$ (the cycloid).

(6) $y^2 = 4ax$.

(7) The ellipse $x = a \cos \theta$, $y = b \sin \theta$.

(8) $x^2 + y^2 + 2Ax + 2By = 0$ at the origin.

(9) $r = a\,(1 + \cos \theta)$.

(10) $x^3 + y^3 = 3axy$ (the Folium) at the origin.

(11) The Limaçon $r = a + b \cos \theta$ at the origin.

(12) $r^n = a^n \cos n\theta$.   (13) $x = 1 - t^2$, $y = t - t^3$.

(14) Prove that $\dfrac{9a}{10}$ is the least value of ρ for the curve

$$x^2 y = a \left\{ x^2 + \frac{a^2}{\sqrt{5}} \right\}.$$

(15) For any plane curve show that

$$\frac{1}{\rho} = \frac{-\dfrac{d^2 x}{ds^2}}{\dfrac{dy}{ds}} = \frac{\dfrac{d^2 y}{ds^2}}{\dfrac{dx}{ds}},$$

and hence that

$$\frac{1}{\rho^2} = \left(\frac{d^2 x}{ds^2} \right)^2 + \left(\frac{d^2 y}{ds^2} \right)^2.$$

(16) Prove also that

$$\frac{1}{\rho} = \frac{\dfrac{1}{r}\left\{\dfrac{d}{ds}\left(r^2\dfrac{d\theta}{ds}\right)\right\}}{\dfrac{dr}{ds}}.$$

(17) If the equation of the curve is given implicitly, e.g. $f(x,y)=0$, prove that

$$\rho = \frac{\left[\left(\dfrac{\partial f}{\partial x}\right)^2 + \left(\dfrac{\partial f}{\partial y}\right)^2\right]^{\frac{3}{2}}}{\dfrac{\partial^2 f}{\partial x^2}\left(\dfrac{\partial f}{\partial y}\right)^2 - 2\dfrac{\partial^2 f}{\partial x\,\partial y}\dfrac{\partial f}{\partial x}\dfrac{\partial f}{\partial y} + \dfrac{\partial^2 f}{\partial y^2}\left(\dfrac{\partial f}{\partial x}\right)^2}.$$

*(18) Show that the origin is an inflexion of the curve

$$(x^2+9y^2)\,x=24y,$$

and give a rough drawing of the curve.

(19) Prove that the curve given by

$$\begin{aligned}x&=at^3+bt^2+ct+d\\y&=At^3+Bt^2+Ct+D\end{aligned}\Bigg\}$$

has in general two inflexions, but only one if $\dfrac{a}{A}=\dfrac{b}{B}$.

What happens if these ratios are equal to $\dfrac{c}{C}$?

(20) Prove that the cubic curve $y=ax^3+bx^2+cx+d$ has one inflexion and find the equation of the tangent there.

§ 89. Contact of Curves.

Suppose that two curves whose equations are $y=\phi(x)$, $y=\psi(x)$, cut at a point P whose coordinates are (x_0, y_0) (Fig. 31).

Now let x_0 be slightly increased to x_0+h, and suppose that this increase brings us to the points P' and P'' on the two curves.

Hence in the figure

$$P'N = \phi(x_0+h),$$
$$P''N = \psi(x_0+h).$$

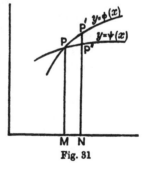

Fig. 31

But, assuming the proper conditions as to continuity and so forth to be satisfied by the two functions $\phi(x)$ and $\psi(x)$, we have by Taylor's Theorem,

$$\phi(x_0 + h) \equiv \phi(x_0) + h\phi'(x_0) + \frac{h^2}{2!}\phi''(x_0) + \dots,$$

$$\psi(x_0 + h) \equiv \psi(x_0) + h\psi'(x_0) + \frac{h^2}{2!}\psi''(x_0) + \dots.$$

Also, since the curves cut at (x_0, y_0), we have

$$\phi(x_0) = \psi(x_0).$$

$$\therefore \quad P'P'' = P'N - P''N$$

$$= h\{\phi'(x_0) - \psi'(x_0)\} + \frac{h^2}{2!}\{\phi''(x_0) - \psi''(x_0)\}$$

$$+ \frac{h^3}{3!}\{\phi'''(x_0) - \psi'''(x_0)\} + \dots \text{ ad inf.}$$

Now let h be a quantity such that its square is negligible. Then a close approximation to $P'P''$ is

$$h\{\phi'(x_0) - \psi'(x_0)\},$$

and if
$$\phi'(x_0) = \psi'(x_0),$$

then P' and P'' coincide and the curves have a second point in common close to P. In fact, they touch in the ordinary geometrical sense of the term, or, more precisely, they have *contact of the first order*.

If further $\phi''(x_0) = \psi''(x_0)$, there will be a third co-incident point on the two curves. There will be a common inflexional tangent and the curves are said to have *contact of the second order*.

And, in general, if the first n derivatives of y with respect to x for both curves are equal for the same value of x, then the curves are said to have at that point *contact of the nth order*.

§ 90. The Circle of Curvature.

Since a circle may always be drawn through any three points (not collinear), we see that, if we take three points P, Q, R as in the figure, and make them ultimately co-incide, the circumcircle of the triangle PQR becomes a circle having three points in common with the curve, i.e. having contact of the second order with the curve at the point P.

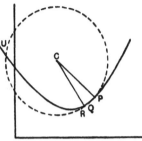

Fig. 32

It is clear that this is the closest contact we can expect between the curve and a circle, and that the centre, being the ultimate position of the intersection of the bisectors of PQ, QR, may be regarded as the intersection of two close normals. It is further clear from the figure that the curve will cut the circle again once at least.

§ 91. The Circle of Curvature (cont.).

We proceed to show that the radius of this " circle of closest contact " is the same as the " radius of curvature " at the point of contact.

Now we have noted above in § 89 that for curves having contact of the second order we may equate the first two differentiations.

Let the equation to the circle be

$$(x - \alpha)^2 + (y - \beta)^2 = R^2.$$

Differentiate with respect to s, the arc of the circle.

$$\therefore \quad (x - \alpha)\frac{dx}{ds} + (y - \beta)\frac{dy}{ds} = 0.$$

But, since first differentiations are equal,

$$\frac{dx}{ds} \text{ for the circle} = \frac{dx}{ds} \text{ for the curve} = \cos \psi,$$

and

$$\frac{dy}{ds} \quad \text{,,} \quad \text{,,} \quad = \frac{dy}{ds} \quad \text{,,} \quad \text{,,} \quad = \sin \psi.$$

Hence $(x - \alpha) \cos \psi + (y - \beta) \sin \psi = 0.$

Differentiate again, this time with respect to ψ, which is the same both for the curve and the circle at the point under consideration.

$$\therefore \quad - (x - \alpha) \sin \psi + \cos \psi \frac{dx}{d\psi} + (y - \beta) \cos \psi$$
$$+ \sin \psi \frac{dy}{d\psi} = 0.$$

$$\therefore \quad (x - \alpha) \sin \psi - (y - \beta) \cos \psi$$
$$= \cos \psi \frac{dx}{d\psi} + \sin \psi \frac{dy}{d\psi}$$
$$= \cos \psi \frac{dx}{ds} \frac{ds}{d\psi} + \sin \psi \frac{dy}{ds} \frac{ds}{d\psi}$$
$$= \frac{ds}{d\psi} \{\cos^2 \psi + \sin^2 \psi\}.$$

We thus have

$$\left. \begin{array}{l} (x - \alpha) \sin \psi - (y - \beta) \cos \psi = \dfrac{ds}{d\psi} \\ (x - \alpha) \cos \psi + (y - \beta) \sin \psi = 0 \end{array} \right\} \ldots\ldots(A).$$

Squaring and adding, we get

$$(x - \alpha)^2 + (y - \beta)^2 = \left(\frac{ds}{d\psi}\right)^2.$$

But $(x - \alpha)^2 + (y - \beta)^2 = R^2.$

Hence $R = \dfrac{ds}{d\psi}.$

Hence the radius of the circle of closest contact is $\dfrac{ds}{d\psi}$ which we have already proved to be ρ, the "radius of curvature" of the curve at the point (x, y) under consideration.

For this reason the circle is known as the "circle of curvature at the given point," and its centre as "the centre of curvature at the given point."

Reverting to equations (A) and solving, we obtain

$$\left. \begin{aligned} x - \alpha &= \frac{ds}{d\psi} \sin \psi = \rho \sin \psi \\ y - \beta &= \qquad\qquad -\rho \cos \psi \end{aligned} \right\},$$

i.e.
$$\left. \begin{aligned} \alpha &= x - \rho \sin \psi \\ \beta &= y + \rho \cos \psi \end{aligned} \right\} \qquad \dots\dots(B).$$

These are two equations giving the coordinates of the centre of curvature. They may also be deduced from the figure by the method of projections. For clearly the centre of curvature at P must lie on the normal at P and at a distance ρ from P.

§ 92. Example.

Find the centre of curvature of the parabola $y^2 = 4ax$ *at any point.*

Using the ordinary parametric notation $x = at^2$, $y = 2at$, the formula (C) of § 87 gives

$$\rho = \frac{(4a^2t^2 + 4a^2)^{\frac{3}{2}}}{-4a^2}$$

$$= -2a(1 + t^2)^{\frac{3}{2}}.$$

(The sign is rightly negative, for as x increases, s increases but ψ decreases, hence $\dfrac{ds}{d\psi}$ is negative.)

Also $$\tan \psi = \frac{1}{t},$$

whence $$\cos \psi = \frac{t}{\sqrt{1 + t^2}}, \quad \sin \psi = \frac{1}{\sqrt{1 + t^2}}.$$

$$\therefore \quad \alpha = x - \rho \sin \psi$$
$$= at^2 + 2a\,(t^2 + 1)^{\frac{3}{2}} \frac{1}{\sqrt{t^2 + 1}} = 2a + 3at^2,$$

and $$\beta = y + \rho \cos \psi$$
$$= 2at - 2a\,(t^2 + 1)^{\frac{3}{2}} \frac{t}{\sqrt{1 + t^2}} = -2at^3.$$

§ 93. The Evolute of a Curve.

Definition. The locus of the centre of curvature of a given curve for all points on it is called the *evolute* of the given curve.

We may thus in the preceding example find the evolute of the parabola $y^2 = 4ax$; for we have found the centre of curvature, which is a typical point on the evolute.

The parametric equation is thus

$$\begin{aligned} x &= 2a + 3at^2 \\ y &= -2at^3 \end{aligned}\Bigg\}.$$

If we desire the Cartesian equation, we eliminate t, getting as a result

$$\left(-\frac{y}{2a}\right)^2 = \left(\frac{x - 2a}{3a}\right)^3,$$

or $$27ay^2 = 4\,(x - 2a)^3.$$

This curve is known as a semi-cubical parabola. We shall however give a simpler method of obtaining the evolute in the next chapter.

EXAMPLES VIII B

(1) Find the radius and centre of curvature at any point of the ellipse $\dfrac{x^2}{a^2}+\dfrac{y^2}{b^2}=1$.

(2) By considering the four intersections of the above ellipse with any circle, write down the coordinates of the other point in which the circle of curvature at "θ" meets the ellipse.

Hence find the centre of curvature.

(3) Apply the method of Example 2 to the parabola $y^2=4ax$.

(4) In the cycloid
$$x=a\,(\theta+\sin\theta),$$
$$y=a\,(1-\cos\theta),$$

prove that
$$\rho=4a\cos\frac{\theta}{2}.$$

Show that the evolute of a cycloid is a similar curve.

*(5) In a plane curve O is the origin, P any point of the curve and C the corresponding centre of curvature. Prove that

$$\cot\angle POC=r\,\frac{d\phi}{dr}\ \text{(usual notation)}.$$

(6) C is the centre of curvature corresponding to any point on the catenary

$$y=\frac{c}{2}\,(e^{\frac{x}{c}}+e^{-\frac{x}{c}}).$$

CP meets OX in G; show that $CP=PG$.

(7) A parabola with its axis parallel to Oy, has closest possible contact with $ay^2=x^3$ at the point (a,a). Find the equation to the parabola.

(8) Find the equation to the circle of curvature at the origin for the curve

$$x^3+y^3+3xy-x+y=0.$$

(9) Find the evolute of the curve

$$x=t+3,\quad y=t^2-4t+6.$$

(10) Find the evolute of the curve

$$x = 3 + \frac{2}{t}, \quad y = 2t - 5.$$

(11) Find the equation of the conic which has the closest possible contact with $ay = x^3$ at the point (a, a).

(12) Prove geometrically that two polar curves $r = \phi(\theta)$, $r = f(\theta)$ will have contact of order n, if the first n derivatives of the functions f and ϕ are equal at a point of intersection of the curves.

(13) Find the polar equation of the circle of closest contact with $r = e^{\theta}$ at a point where $\theta = \frac{\pi}{2}$.

(14) Use this method to find the equation to the tangent to the polar curve

$$r^2 = a^2 \cos 2\theta \text{ at a point where } \theta = \frac{\pi}{6}.$$

N.B. This method of finding the line of closest contact is often the best way of obtaining the tangent to a curve in polars.

(15) Use the method of Example 2 to find the eccentric angle of any point on the ellipse, such that the circle of curvature there meets the ellipse again in a fixed point "α." Prove that three such points exist; and also that they are concyclic with the point "α."

(16) The circle of curvature at a point P of a conic meets the conic again in Q. Prove that the "chord of curvature"* PQ and the tangent at P are equally inclined to the axis of the conic.

(17) If ρ_1 and ρ_2 are the radii of curvature at the ends of a focal chord of $y^2 = 4ax$, prove that $\rho_1{}^{-\frac{2}{3}} + \rho_2{}^{-\frac{2}{3}} = (2a)^{-\frac{2}{3}}$.

(18) Find the equation to the "chord of curvature" at $\left(ct, \dfrac{c}{t} \right)$ on the hyperbola $xy = c^2$.

---

* Many writers call this chord "the chord of curvature" at P. Some, however, use the expression "a chord of curvature at P in a given direction" to denote any chord of the circle of curvature drawn through P.

CHAPTER IX

ENVELOPES AND ASSOCIATED LOCI

§ 94. The Envelope of a Family of Lines.

We often need to find the curve to which each of a given set of lines is a tangent; or, more generally, to find a curve which is touched by each of a given set of curves. This is known as finding the " envelope " of the given set, or " family," of lines or curves.

It is clear that, in order to possess an envelope at all, the family of lines or curves must all be expressed in terms of one variable parameter only. [For if they are all tangents to a curve, they will be expressed in terms of the parameter appropriate to that curve.]

The procedure is best explained by an example.

Find the envelope of the family of straight lines

$$y = mx - 2am - am^3,$$

where m is a variable parameter.

The line $y = mx - 2am - am^3$ is a tangent to the envelope for all values of m, and therefore in particular for $m + \delta m$ as well as for m.

Hence

$$\left. \begin{aligned} y &= mx - 2am - am^3 \\ y &= (m + \delta m)\, x - 2a\, (m + \delta m) - a\, (m + \delta m)^3 \end{aligned} \right\}$$

are the equations of two close tangents to the envelope.

Subtracting and dividing by δm, we see that

$$\text{and} \quad \left. \begin{aligned} y &= mx - 2am - am^3 \\ 0 &= x - 2a - a\, (3m^2 + 3m\, \delta m + \delta m^3) \end{aligned} \right\}$$

are two equations which are true for the intersection of two close tangents.

Next, let $\delta m \to 0$, then the two tangents coincide and their intersection becomes the point of contact of either of them with the envelope.

Thus the two equations

$$\left. \begin{array}{l} y = mx - 2am - am^3 \\ 0 = x - 2a - 3am^2 \end{array} \right\}$$

are typical of the point of contact of the line with the envelope, and if we eliminate m between them, we obtain the locus of this point of contact, which is of course the envelope itself.

We see that what we have really done is to differentiate the given equation with respect to m, treating x and y as constants; and we also observe that the argument does not in any way restrict the form of the original equation to be that of a straight line, but that the argument is equally true for a family of curves.

Hence the rule is as follows:

To find the envelope of a family of lines or curves given by ϕ (x, y, c) = 0 for different values of c, write down the two equations

$$\left. \begin{array}{l} \phi \text{ (x, y, c) = 0} \\ \dfrac{\partial}{\partial c}\, \phi \text{ (x, y, c) = 0} \end{array} \right\}$$

and then eliminate c. The result is, or at any rate contains, the equation to the envelope.

The qualification contained in the last sentence is due to the fact that it can be proved that this eliminant may also contain other factors, such as the loci of cusps, nodes, and so on. These extraneous factors are, however, easily detected by the fact that they are repeated (either as squares or cubes).

A similar result holds for the family of curves

$$\phi (r, \theta, c) = 0.$$

§ 95. Two Connected Parameters.

When the line or curve whose envelope is to be found involves two parameters connected by another relation, a method analogous to that considered for the parallel case in Maxima and Minima (cf. § 47) may be employed. This method is illustrated by the following problem.

Find the envelope of a line which includes with the axes a triangle of constant area c^2.

Using the intercept form, we have to find the envelope of

$$\frac{x}{a} + \frac{y}{b} = 1 \quad \dots\dots\dots\dots\dots(1),$$

subject to the condition

$$ab = 2c^2 \quad \dots\dots\dots\dots\dots(2).$$

Consider that a and b are both functions of some variable t; differentiate with respect to t, keeping x and y constant. Then

$$\left. \begin{array}{r} -\dfrac{x}{a^2}\dfrac{da}{dt} - \dfrac{y}{b^2}\dfrac{db}{dt} = 0 \\[2mm] b\dfrac{da}{dt} + a\dfrac{db}{dt} = 0 \end{array} \right\}.$$

Eliminating $\dfrac{da}{dt}$ and $\dfrac{db}{dt}$ we get, for any point on the envelope,

$$\frac{x}{a^2 b} = \frac{y}{b^2 a}.$$

$$\therefore \frac{x}{a} = \frac{y}{b} = \frac{1}{2},$$

since
$$\frac{x}{a} + \frac{y}{b} = 1.$$

$$\therefore a = 2x, \quad b = 2y.$$

But
$$ab = 2c^2.$$

$$\therefore 4xy = 2c^2$$

for any point on the envelope.

Hence the envelope is the rectangular hyperbola

$$xy = \frac{c^2}{2},$$

having the axes as asymptotes.

§ 96. Trigonometrical Equations.

We append an example showing how to deal with equations containing trigonometrical functions.

Find the envelope of $x\sqrt{\cos\theta} + y\sqrt{\sin\theta} = a$ for different values of θ.

Differentiating with respect to θ, we have

$$\tfrac{1}{2}x\,\frac{-\sin\theta}{\sqrt{\cos\theta}} + \tfrac{1}{2}y\,\frac{\cos\theta}{\sqrt{\sin\theta}} = 0.$$

$$\therefore x\sin^{\frac{3}{2}}\theta = y\cos^{\frac{3}{2}}\theta,$$

i.e.
$$\left.\begin{array}{l} \sin^{\frac{3}{2}}\theta = \lambda y \\ \cos^{\frac{3}{2}}\theta = \lambda x \end{array}\right\} \quad \dots\dots\dots\dots\dots(A)$$

or
$$\sin\theta = \lambda^{\frac{2}{3}} y^{\frac{2}{3}}, \quad \cos\theta = \lambda^{\frac{2}{3}} x^{\frac{2}{3}}.$$

But since
$$\sin^2\theta + \cos^2\theta = 1,$$

we have
$$\lambda^{\frac{4}{3}} = \frac{1}{x^{\frac{4}{3}} + y^{\frac{4}{3}}}.$$

Reverting to (A)

$$\sin \theta = \frac{y^{\frac{2}{3}}}{\sqrt{x^{\frac{4}{3}} + y^{\frac{4}{3}}}} \Bigg\} .$$

$$\cos \theta = \frac{x^{\frac{2}{3}}}{\sqrt{x^{\frac{4}{3}} + y^{\frac{4}{3}}}} \Bigg)$$

Substituting in the original equation we get

$$\frac{x^{\frac{4}{3}}}{\sqrt[4]{x^{\frac{4}{3}} + y^{\frac{4}{3}}}} + \frac{y^{\frac{4}{3}}}{\sqrt[4]{x^{\frac{4}{3}} + y^{\frac{4}{3}}}} = a;$$

i.e. $\qquad\qquad \{x^{\frac{4}{3}} + y^{\frac{4}{3}}\}^{\frac{3}{4}} = a.$

Hence the equation to the envelope is $x^{\frac{4}{3}} + y^{\frac{4}{3}} = a^{\frac{4}{3}}.$

§ 97. "The Tangential Equation" of a Curve.

In Analytical Geometry it is often required to find the condition under which the line $lx + my + 1 = 0$ {or sometimes $lx + my + n = 0$} touches a given conic or a given curve. The resulting condition is often known as the **Tangential Equation** of the conic or curve. A method of finding the tangential equation of a curve is illustrated by the following example.

Given that $lx + my + 1 = 0$ touches the curve $x^3 + y^3 + c^3 = 0$, find the relation between l and m.

Consider x and y as functions of some parameter t, and differentiate both equations with respect to t.

$$\therefore \quad 3x^2 \frac{dx}{dt} + 3y^2 \frac{dy}{dt} = 0 \Bigg\} .$$

$$l \frac{dx}{dt} + m \frac{dy}{dt} = 0 \Bigg)$$

Therefore eliminating $\dfrac{dx}{dt}$ and $\dfrac{dy}{dt}$,

$$\frac{l}{x^2} = \frac{m}{y^2}.$$

Hence
$$l = \lambda x^2 \atop m = \lambda y^2 \Big\}.$$

But since $lx + my + 1 = 0$,
$$\lambda = -\frac{1}{x^3 + y^3} = \frac{1}{c^3}.$$

Hence
$$x^2 = c^3 l,$$
$$y^2 = c^3 m.$$
$$\therefore \quad x = (c^3 l)^{\frac{1}{2}}, \quad y = (c^3 m)^{\frac{1}{2}}.$$

But
$$x^3 + y^3 = c^3.$$
$$\therefore \quad (c^3 l)^{\frac{3}{2}} + (c^3 m)^{\frac{3}{2}} = c^3.$$

Hence the condition is
$$c^{\frac{3}{2}} \{ l^{\frac{3}{2}} + m^{\frac{3}{2}} \} = 1.$$

§ 98. The Evolute considered as an Envelope.

We have mentioned in § 90 that the centre of curvature of a curve may be regarded as the intersection of two close normals to the curve, or, to put the matter more exactly, the limiting position taken up by the intersection of two normals when the normals themselves ultimately coincide. It also follows from this argument that every normal touches the evolute; for, if we consider three normals, a, b, c, such that the two outer ones a, c are both tending to coincidence with b, the intersections of a, b, and of b, c ultimately become points on the evolute, and in the limit coincide. Therefore the normal b contains two coincident points of the evolute and thus touches it. It follows then that the evolute can be obtained by finding the envelope of the normals; and this method is frequently the best way of finding both the evolute and the centre of curvature, which is the point of contact of the evolute with the normal.

Reverting to the same example we tried in § 93, viz. the evolute of $y^2 = 4ax$, we see that the work can be arranged as follows.

The normal to the parabola at $(at^2,\ 2at)$ is

$$y + tx - 2at - at^3 = 0 \ \dots\dots\dots\dots(i).$$

To find the envelope differentiate with respect to t.

$$\therefore \quad x - 2a - 3at^2 = 0 \ \dots\dots\dots\dots(ii).$$

Therefore the point of contact has the abscissa $2a + 3at^2$, and the ordinate can be found at once to be $-2at^3$.

Hence the point of contact, i.e. the centre of curvature, is $(2a + 3at^2,\ -2at^3)$.

We now proceed, as before, to eliminate t between

$$\left.\begin{array}{l} x = 2a + 3at^2 \\ y = -2at^3 \end{array}\right\},$$

and the result is the equation $27ay^2 = 4(x - 2a)^3$. If the centre of curvature is not required, we can eliminate t immediately between equations (i) and (ii) without finding x and y.

§ 99. Associated Loci. The First Positive Pedal.

The locus of the foot of the perpendicular from the origin on any tangent to a curve is called the *First Positive Pedal* of the curve with respect to the origin.

The method of finding the first positive pedal is easily seen. We find the equation to any tangent to the curve and rewrite it in the form $x \cos \alpha + y \sin \alpha = p$.

But p and α are the polar coordinates of the foot of the perpendicular from O on the tangent, and, performing the necessary eliminations, we get a relation between p and α which is the polar equation of the First Positive Pedal.

Parametric notation should be used wherever possible. The method is illustrated by the following example.

Find the First Positive Pedal of the ellipse

$$\frac{x^2}{a^2} + \frac{y^2}{b^2} = 1$$

with respect to its centre.

The equation of any tangent is

$$\frac{x}{a} \cos \theta + \frac{y}{b} \sin \theta = 1.$$

Comparing this with $x \cos \alpha + y \sin \alpha = p$ we have

$$\frac{\cos \theta}{a \cos \alpha} = \frac{\sin \theta}{b \sin \alpha} = \frac{1}{p};$$

i.e.
$$\cos \theta = \frac{a \cos \alpha}{p}; \quad \sin \theta = \frac{b \sin \alpha}{p}.$$

$$\therefore \quad a^2 \cos^2 \alpha + b^2 \sin^2 \alpha = p^2.$$

Now p and α are polar coordinates of the foot of the perpendicular; hence, replacing p and α by r and θ, we get the polar equation of the first positive pedal, viz.

$$r^2 = a^2 \cos^2 \theta + b^2 \sin^2 \theta.$$

{In Cartesian coordinates $(x^2 + y^2)^2 = a^2 x^2 + b^2 y^2$.}

A similar method can easily be devised for curves whose equations are originally in Polar Coordinates.

§ 100. Inversion.

If a fixed point O be joined to any point P on a given curve and another point P' taken on OP such that $OP . OP' = k^2$, where k is a constant, then, as P moves round the curve, P' traces out a locus which is known as the **inverse of the given curve with respect to O.** O is known as the **centre,** and k as the **radius, of inversion.**

Inversion is largely treated in treatises on Pure Geometry, but the subject may easily be approached analytically as follows:

Transform the equation of the given curve into Polar Coordinates, which are obviously best suited for this calculation. Then if P is (r, θ), P' must be $\left(\dfrac{k^2}{r}, \theta\right)$. Hence, if the equation to the original curve is $f(r, \theta) = 0$, the equation to the inverse curve will be $f\left(\dfrac{k^2}{r}, \theta\right) = 0$.

Example. Find the inverse of a parabola with respect to its focus.

The polar equation of the parabola with respect to its focus is

$$\frac{2a}{r} = 1 - \cos \theta.$$

Then, writing $\dfrac{k^2}{r}$ for r, the polar equation of the inverse curve is

$$\frac{2ar}{k^2} = (1 - \cos \theta)$$

or

$$r = \frac{k^2}{2a}(1 - \cos \theta),$$

which is the equation to a cardioid.

A third class of derived locus is the " Polar Reciprocal," which is the inverse of the First Positive Pedal: but this curve is seldom required for curves other than conics when it can be derived more easily otherwise.

EXAMPLES IX

Find the envelopes of the following lines and curves:

(1) $y = mx + am^2$.   (2) $x \cos \theta + y \sin \theta = a$.

(3) $x \cos^3 \theta + y \sin^3 \theta = a$.   (4) $x \cos 2\theta + y \sin 2\theta = a \cos^2 \theta$.

(5) $\dfrac{x}{a} + \dfrac{y}{b} = 1$, given that $a + b = c$.

(6) The same line, given that $a^2 + b^2 = c^2$.

(7) The ellipse $\dfrac{x^2}{a^2} + \dfrac{y^2}{b^2} = 1$, given that $a^2 + b^2 = c^2$.

(8) *The family of parabolas $y = x \tan \theta - \dfrac{1}{2} \dfrac{gx^2}{a^2 \cos^2 \theta}$.

(9) Restate the results of Examples 5 and 7 in geometrical language, without using symbols.

Find the "tangential" equations of the following:

(10) The circle $x^2 + y^2 = a^2$.   (11) The parabola $y^2 = 4ax$.

(12) The semicubical parabola $ay^2 = x^3$.

(13) The hyperbola $xy = c^2$.   (14) $\dfrac{1}{x} + \dfrac{1}{y} + \dfrac{1}{a} = 0$.

(15) $\sqrt{\dfrac{x}{a}} + \sqrt{\dfrac{y}{b}} = 1$.

(16) From any point on the ellipse $\dfrac{x^2}{a^2} + \dfrac{y^2}{b^2} = 1$ perpendiculars are drawn to the axes. Find the envelope of the line joining their feet.

Find the evolutes of the following curves, giving the coordinates of the centre of curvature where possible:

(17) The ellipse $\dfrac{x^2}{a^2} + \dfrac{y^2}{b^2} = 1$.   (18) $xy = c^2$.

(19) $x = 1 - t^2$, $y = 4 - t$.   (20) $x = 1 - \cos t$, $y = t - \sin t$.

* These parabolas are the trajectories of a particle projected at different angles of projection with given velocity.

(21) Prove that in polar coordinates the first positive pedal of the curve $f(r, \theta) = 0$ is the relation between R and Θ obtained by elimination between

$$\left. \begin{aligned} R &= r \sin \phi \\ \Theta &= \theta + \phi - \frac{\pi}{2} \end{aligned} \right\},$$

ϕ having its usual significance.

(22) Prove that the first positive pedal of a circle with respect to any point is a Limaçon $(r = a + b \cos \theta)$.

(23) Find the first positive pedal of the parabola with respect to (1) the focus, (2) the vertex.

(24) Find the first positive pedal of the ellipse with respect to a focus and express the result as a geometrical theorem.

(25) Prove that the inverse of a curve $f(x, y) = 0$ is the curve

$$f \left\{ \frac{k^2 x}{x^2 + y^2}, \ \frac{k^2 y}{x^2 + y^2} \right\} = 0.$$

(26) Prove that the inverse of a circle with respect to any point is a circle, except when the point lies on the circle, when the inverse degenerates into a straight line.

(27) Coplanar rays of light parallel to the axis of x fall upon a concave mirror in the form of half the circle $x^2 + y^2 = a^2$ (1st and 4th quadrants).

Find the equation of the reflected ray at the point $(a \cos \theta, a \sin \theta)$.

Find the envelope of these reflected rays. (This curve is known as the "caustic" of the mirror.)

(28) Rays parallel to the axis of y fall on a concave mirror in the form of the cycloid

$$x = a(\theta + \sin \theta),$$
$$y = a(1 - \cos \theta).$$

Find the equation to the "caustic."

CHAPTER X

SIMPLE CURVE TRACING

§ 101. General Ideas.

It is assumed that the reader is familiar with certain general ideas on the subject of the tracing of curves whose equations are given. For convenience of reference we collect these results below.

(1) *If the equation is unaltered when x is changed into − x, then the curve is symmetrical with respect to the axis of y.*

(2) *Similarly, if y can be changed into −y, without affecting the equation, then the curve is symmetrical with respect to the axis of x.*

(3) *So also, if y and x can be interchanged, the curve is symmetrical about the line y = x; and if y and − x can be changed, it is symmetrical about y + x = 0.*

(4) *A curve of degree n cannot be met by a straight line in more than n points.*

The first important step in the tracing of a curve is usually to determine its form at the origin. If the origin is not on the curve in the form in which it is given, the origin can always be changed so that some known point on the curve becomes the origin. We now propose to show how, by rejecting some of the terms and retaining others, some idea of the shape near the origin may be formed. In what follows, the student should observe that the object is rather to ascertain the nature of the curve than its precise position. The latter can only be done after the former, and then only by a laborious process of plotting.

§ 102. The Tangent at the Origin.

The Cartesian equation of any algebraic curve may be written in a number of sets of homogeneous terms, thus

$$u \equiv a_0 + (a_1 x + b_1 y) + (a_2 x^2 + b_2 xy + c_2 y^2) + \dots$$
$$+ (a_n x^n + b_n x^{n-1} y + \dots) = 0.$$

Since we are considering a curve which passes through the origin, we may take a_0 to be zero.

We now proceed to show that the tangent (or tangents) at O can be found by equating to zero the lowest set of homogeneous terms. For, if we transform the equation into polar coordinates, it becomes

$$r (a_1 \cos \theta + b_1 \sin \theta)$$
$$+ r^2 (a_2 \cos^2 \theta + b_2 \cos \theta \sin \theta + c_2 \sin^2 \theta) + \dots$$
$$+ r^n (a_n \cos^n \theta + \dots) = 0.$$

If in this equation we give θ any particular value, say $\theta = \frac{\pi}{4}$, we get an equation of the nth degree in r giving the n different distances from O at which the curve is cut by a line through O in the particular direction chosen. Also, since r is a factor of the whole equation, one of these distances will be zero, which is only what we should expect from the fact that the curve goes through O.

But if, instead of $\theta = \frac{\pi}{4}$, we choose the particular value of θ given by $a_1 \cos \theta + b_1 \sin \theta = 0$, then the first term disappears and r^2 becomes a factor of the resulting equation. Hence the direction given by $a_1 \cos \theta + b_1 \sin \theta = 0$ is one which cuts the curve twice at the origin, i.e. is the direction of the tangent at the origin.

But the equation $a_1 \cos \theta + b_1 \sin \theta = 0$, when converted back into Cartesian coordinates, becomes $a_1 x + b_1 y = 0$, which is thus the tangent at the origin.

If this term happens to be missing in the equation, the next term will be of the second degree and will probably factorize into a pair of tangents, and so forth.

In general, the lowest homogeneous set of terms can be factorized into (a) linear factors rational or irrational, (b) (possibly) some irreducible quadratic factors. Each of the linear factors (a) will give us a real tangent to the curve, but each of the factors (b) will give us no real tangents, i.e. there is no real branch of the curve belonging to these factors. If only factors such as (b) exist, the origin is said to be an "isolated point" or "acnode" on the curve. An irreducible factor of quite frequent occurrence is the factor $(x^2 + y^2)$.

If there are two or more real tangents the origin is called a "double" or "multiple" point of the curve. Various types of multiple point can occur, e.g. nodes, cusps, etc., which we shall illustrate later (Fig. 33).

§ 103. Branches at the Origin.

Corresponding to every tangent at O, we have a branch touching the tangent at O. Now if $x = 0$ is the tangent, the branch must in the final resort reduce down to $x = Ay^{r+1}$, where r is some positive number. This is seen to be the case since $x = 0$ has to be the tangent, and other powers of x become negligible, for, in approaching the origin, x becomes very small while y still remains of appreciable size (Fig. 33a).

Similarly, a branch touching $y = 0$ will be reduced finally to $y = Bx^{r+1}$ for similar reasons (Fig. 33b).

Lastly, suppose $y - mx = 0$ is a tangent. Then it is seen from Fig. 33c that, as we approach the origin, $(y - mx)$ becomes smaller than either x or y. Therefore all higher powers of $(y - mx)$ disappear, and the branch can now be expressed in either of the forms

$$y - mx = Ax^{r+1}$$

or $$y - mx = By^{s+1},$$

of which we usually choose the first form merely as a convention.

We see that the final types to be considered are three, viz.

$$x = Ay^{r+1},$$
$$y = Bx^{r+1},$$
$$y - mx = Ax^{r+1},$$

of which the first two may be called "axial," and the last "oblique," branches.

§ 104. Axial Branches.

We will consider first how to draw axial branches when we have obtained their equations. A few examples will make clear the method to be adopted.

(1) $x^3 = -k^2 y$.

Now $y = 0$ is the tangent, and to make x real y must always be negative (Fig. 33d).

(2) $x^4 = k^2 y^7$.

Here $x = 0$ is the tangent and y must be positive. The branch is a "keratoid" or "horn-shaped" cusp (Fig. 33e). The other type of cusp, viz. a "rhamphoid" or "beak-shaped" cusp, is shown in Fig. 33f.

(3) $x^5 = -k^3 y^7$.

Here $x = 0$ is the tangent and x and y are of opposite signs. The branch is an "inflexion" (Fig. 33g).

(4) $y^4 = k^4 x^6$.

This factorizes into $y^2 = k^2 x^3$, $y^2 = -k^2 x^3$, giving us two branches as in Fig. 33h.

§ 105. Oblique Branches.

To trace the oblique branches $y = mx + Ax^{n+1}$ we proceed as follows.

(1) $y - x = -k^2 x^2$.

$y = x$ is the tangent, and since the "correction" $-k^2 x^2$ is always negative whatever the sign of x, y must be less than x and thus the curve is below its tangent (Fig. 33k).

(2) The "correction" may not be single-valued, e.g. $(y + x)^2 = -k^2 x^7$.

Here $y + x = 0$ is the tangent and $y + x = \pm \sqrt{-k^2 x^7}$. Hence x must be negative and the curve is both below and above the tangent (Fig. 33l).

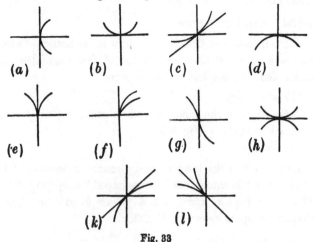

Fig. 33

§ 106. Methods of obtaining Equations to Branches.

There are three methods by which we can choose the terms which are to be retained as the equation to a branch at O. They are

(1) the homogeneous reduction,

(2) the division reduction,

(3) the graphic reduction.

The three methods should be tried in this order, but in many curves the graphic reduction will not be required.

§ 107. The Homogeneous Reduction.

In finding a branch touching $x = 0$, we may put $x = 0$ in every factor of a homogeneous set of terms, except in the factor x itself (or a power of x).

In general, that is, we may replace a set of terms

$$x^a y^b (y - mx)^c (y - nx)^d \dots$$

by
$$x^a y^{b+c+d}.$$

For we have said that x decreases more rapidly than y as we approach O. Hence, within the set of terms itself, we may neglect x in comparison with y. But we clearly must not do this in the term x^a, for that would destroy the whole set of terms.

Similarly, in finding a branch touching $y = 0$, we put $y = 0$ in every factor of a set except in the factor y itself.

Again, for branches touching $y = mx$, we put $y = mx$, except in the factor $(y - mx)$ itself.

We thus see that the effect of the homogeneous reduction is to reduce each set of terms to a single term, but that no set of terms can disappear completely.

§ 108. The Division Reduction.

In finding branches touching either of the axes we may neglect a term which is divisible by any other term. For, after performing the homogeneous reduction, we are left with a number of terms $A x^a y^b$, $A' x^{a+a'} y^{b+b'}$, ... and, since for an axial branch either x or y is decreasing very rapidly, the set containing the higher powers may clearly be neglected. [On the other hand, if we have two terms such as $x^2 y^3$, $x^3 y^2$, which are not divisible by one another, it is not quite clear which is the smaller and so both must be left.]

Similarly, for an oblique branch touching $y = mx$, we put $y = mx$, but write β for the factor $(y - mx)$ itself, finally arriving at sets of terms like $A x^a \beta^b$, $A' x^{a+a'} \beta^{b+b'}$, and so on; and here again we can reject the second term in comparison with the first.

These two reductions are often sufficient to determine the form of the curve at O, as the following three examples will show. *But we must always do the homogeneous reduction first.*

§ 109. Worked Examples.

(i) $$x^3 + y^3 - 3axy = 0.$$
The two tangents at O are $x = 0$, $y = 0$.

For $x = 0$. Put $x = 0$, except in the factor x.
$$\therefore \quad y^3 - 3axy = 0$$
or $$y^2 = 3ax.$$
This gives branch α in Fig. 34a.

For $y = 0$. Put $y = 0$, except in the factor y.
$$\therefore \quad x^3 - 3axy = 0$$
or $$x^2 = 3ay,$$
giving the branch β in the same figure.

Hence altogether the form is as in Fig. $34a$; and the origin is a "node" or "double point" of the curve.

(ii) $\qquad x^5 + 2y^5 - 4x^4 - 4x^2y^2 + 3xy^2 = 0,$

i.e. $\qquad x^5 + 2y^5 - 4x^2(x^2 + y^2) + 3xy^2 = 0.$

The tangents are $x = 0,\ y = 0.$

For $x = 0$. Put $x = 0$, except in the factor x.

$$\therefore\quad 2y^5 - 4x^2y^2 + 3xy^2 = 0.$$

Neglect $-4x^2y^2$, since x^2y^2 is divisible by xy^2.

$$\therefore\quad 2y^5 + 3xy^2 = 0$$

or $\qquad\qquad x = -\tfrac{2}{3}y^3,$

giving branch α in Fig. $34b$.

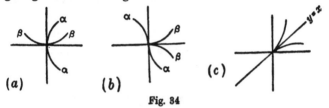

Fig. 34

For $y = 0$. Put $y = 0$, except in the factor y.

$$\therefore\quad x^5 - 4x^4 + 3xy^2 = 0.$$

Omit x^5, divisible by x^4.

$$\therefore\quad -4x^4 + 3xy^2 = 0$$

or $\qquad\qquad y^2 = \tfrac{4}{3}x^3,$

branch β in Fig. $34b$.

Hence the whole shape is as Fig. $34b$.

(iii) $\quad (y^2 - x^2)^2 - 2a(x^3 + y^3) + a^2(y - x)^2 = 0$

or

$(y - x)^2(y + x)^2 - 2a(x + y)(x^2 - xy + y^2) + a^2(y - x)^2 = 0.$

The only tangent is $y = x$.

Put $y = x$, but $y - x = \beta$.

$$\therefore \quad 4\beta^2 x^2 - 4ax^3 + a^2\beta^2 = 0.$$

Omit the term $\beta^2 x^2$, divisible by β^2.

$$\therefore \quad a^2\beta^2 = 4ax^3.$$

$$\therefore \quad \beta = \pm \sqrt{\frac{4x^3}{a}}$$

and the branch is $y = x \pm \sqrt{\dfrac{4x^3}{a}}$.   (Fig. 34$c$.)

§ 110. The Graphic Reduction or " Triangle."

If these two reductions leave us in the end more than two terms, it is necessary to use "de Gua's Triangle." But as this method is somewhat beyond the scope of this work, we give no account of it here. The reader desirous of further information on this interesting method will find it in Hilton's *Plane Algebraic Curves*, p. 38.

EXAMPLES X A

Trace the following branches at O:

(1) $y^2 = k^2 x$, $y^4 = -k^3 x$, $x^5 = k^2 y$.

(2) $y^2 = k^2 x^3$, $y^4 = -k^2 x^5$, $y = -k^2 x^5$.

(3) $y^2 = k^4 x^4$, $y^2 = k^4 x^6$.

(4) $y = x + k^2 x^3$, $y = -x - k^2 x^3$, $(y - x)^4 = k^2 x^5$.

Find the form near O of the following :

(5) $x^3(y^2 - a^2) = y^3(x^2 - 8a^2)$.   (6) $a(x - y) + x^3 + y^3 = 0$.

(7) $(x^2 + y^2)(3x + 4y) = 3x^2 - 10xy + 3y^2$.

(8) $x^4 = (x^2 - y^2)(x - 2y)$.

(9) $(y + 3x)^2(y^2 - x^2) + 3a(y + 3x)x^2 - 5a^2 x^2 = 0$.

(10) $2x(x - y)^2 - 3a(x^3 - y^2) + 4a^2 y = 0$.

(11) $x(x + y)^2 = a(x - y)^2$.   (12) $2xy(x - y)^2 - x^4(x - y) - y^6 = 0$.

(13) $x^5 + y^5 = 5x^2y^2$.   (14) $(y - x^2)^2 = 2x^5 - x^6$.

(15) $x^3 = y(x - a)^2$.   (16) $y^5 + 2xy^2(x+y) + x^2(x+y) = 0$.

(17) By transforming to $(a, 0)$ and $(0, a)$ determine the shape of the curve $\sqrt{x} + \sqrt{y} = \sqrt{a}$.

(18) Show that the curve $x^{\frac{2}{3}} + y^{\frac{2}{3}} = a^{\frac{2}{3}}$ has four cusps on the axes of coordinates.

(19) Transform the equation to the cardioid $r = a(1 + \cos\theta)$ to Cartesians and verify its shape at O.

(20) If u_n denotes a homogeneous set of terms of degree n, examine the nature near O of the following:

         (i) $u_1 + u_1v_1 + u_3 = 0$.

         (ii) $u_1^2 + u_1u_2 + u_4 = 0$.

         (iii) $u_1v_1 + u_1v_2 + u_4 + \ldots = 0$.

(21) Investigate the shape of the curve

$$2x^3y^3 + x^4 - y^4 - 2xy = 0$$

at O and where $x = \pm 1$, $y = \pm 1$.

(22) Transforming the equation $r = a + b\cos\theta$ to Cartesian coordinates, obtain the form at the three points in which it cuts the axis of x.

§ 111. Singularities.

We append a few differential tests which are sometimes of use. They usually enable us to detect the existence of peculiarities that are suspected, rather than to trace the curve in the neighbourhood of those peculiarities; and, in general, it is often no more difficult to find the form of the curve by the preceding methods than to remember and quote the appropriate differential test.

(i) Double points.

The double and multiple points are all roots of

$$\frac{\partial u}{\partial x} = \frac{\partial u}{\partial y} = 0.$$

For
$$\frac{dy}{dx} = -\frac{\dfrac{\partial u}{\partial x}}{\dfrac{\partial u}{\partial y}},$$

and at a double point it is essential that $\dfrac{dy}{dx}$ should have different values corresponding to the branch along which we approach the point. Hence the fraction $\dfrac{\dfrac{\partial u}{\partial x}}{\dfrac{\partial u}{\partial y}}$ must be indeterminate. Now in an algebraical curve neither $\dfrac{\partial u}{\partial x}$ nor $\dfrac{\partial u}{\partial y}$ will tend to infinity, hence both must be zero.

(ii) **Cusps.**

Again, let (X, Y) be such a point and let the origin be transferred to it.

The equation $u(x, y) = 0$ now becomes
$$u(x + X, \ y + Y) = 0,$$
which can be expanded by Taylor's Theorem (Extended Form) (see Ch. XII, § 133).

It becomes
$$u(X, Y) + x\frac{\partial U}{\partial X} + y\frac{\partial U}{\partial Y}{}^{*}$$
$$+ \left(x^2\frac{\partial^2 U}{\partial X^2} + 2xy\frac{\partial^2 U}{\partial X \partial Y} + y^2\frac{\partial^2 U}{\partial Y^2}\right) + \ldots = 0.$$

Now, by hypothesis,
$$u(X, Y) \ \text{and} \ \frac{\partial U}{\partial X} \ \text{and} \ \frac{\partial U}{\partial Y}$$
are zero, since (X, Y) is a double point on the curve.

---

  * $\dfrac{\partial U}{\partial X}$ denotes the value of $\dfrac{\partial u}{\partial x}$ when x, y are replaced by X, Y.

Therefore the lowest terms are now

$$x^2 \frac{\partial^2 U}{\partial X^2} + 2xy \frac{\partial^2 U}{\partial X \partial Y} + y^2 \frac{\partial^2 U}{\partial Y^2} = 0.$$

Hence the tangents at the double point are coincident if

$$\left(\frac{\partial^2 U}{\partial X \partial Y}\right)^2 = \left(\frac{\partial^2 U}{\partial X^2}\right)\left(\frac{\partial^2 U}{\partial Y^2}\right).$$

But a double point with coincident tangents is a cusp.

Hence the full conditions for a cusp are

$$\frac{\partial u}{\partial x} = 0, \quad \frac{\partial u}{\partial y} = 0 \quad and \quad \left(\frac{\partial^2 u}{\partial x \partial y}\right)^2 = \left(\frac{\partial^2 u}{\partial x^2}\right)\left(\frac{\partial^2 u}{\partial y^2}\right).$$

In the same way the tangents will be imaginary if

$$\left(\frac{\partial^2 u}{\partial x \partial y}\right)^2 < \left(\frac{\partial^2 u}{\partial x^2}\right)\left(\frac{\partial^2 u}{\partial y^2}\right).$$

Hence this, in conjunction with $\frac{\partial u}{\partial x} = 0$, and $\frac{\partial u}{\partial y} = 0$, is the condition for an isolated point or " acnode."

§ 112. Inflexions.

We have noticed various conditions for points of inflexion in the chapter on Curvature, the most useful for algebraic curves being

$$\frac{\partial^2 u}{\partial x^2}\left(\frac{\partial u}{\partial y}\right)^2 - 2\frac{\partial^2 u}{\partial x \partial y}\frac{\partial u}{\partial x}\frac{\partial u}{\partial y} + \frac{\partial^2 u}{\partial y^2}\left(\frac{\partial u}{\partial x}\right)^2 = 0^*.$$

An inflexion is not, of course, a double or multiple point, but merely a point where the curve crosses its tangent. It is, however, usually reckoned among the singularities of the curve.

§ 113. Singularities in Parametric Coordinates.

Double points of curves given parametrically can often be found by investigating whether the same values of x and y are given by two different values of t.

* See Ch. xii, § 131, for proof.

If so, then the values of $\frac{dy}{dx}$ may be found; if they are different, the point is a node; but if they are the same, then it is a cusp. Or again, some value of t may be found for which $\frac{dy}{dx}$ is indeterminate; this also will usually give a cusp.

The condition for an inflexion is obtained by making $\rho \to \infty$, whence we usually get $\frac{d^2y}{dt^2}\frac{dx}{dt} = \frac{d^2x}{dt^2}\frac{dy}{dt}$ (§ 87)*. An acnode will result if any imaginary value of t gives a real value to x and y.

Examples. (i) $\qquad \left. \begin{aligned} x &= t^4 - 2t^2 \\ y &= t^3 - 3t \end{aligned} \right\}$.

The point (3, 0) is given by $t = \pm \sqrt{3}$.

But $\qquad \dfrac{dy}{dx} = \dfrac{3(t^2-1)}{4t(t^2-1)} = \dfrac{3}{4t}$,

giving two differing values, and thus the point is a node.

But if $t = \pm 1$, $\frac{dy}{dx}$ is of the form $\frac{0}{0}$, but tends to a single* limiting value $\frac{3}{4}$ (or $-\frac{3}{4}$ in the negative case). These points are cusps.

(ii) The forms of parametric curves at O can be found as follows.

Using the same example, the origin is given only by $t = 0$, and clearly $x \to 0$ first. Hence $x = 0$ is the tangent.

As $t \to 0$ the appreciable terms are

$$x = -2t^2, \quad y = -3t.$$

Therefore, eliminating, the form is $2y^2 + 9x = 0$.

---

* In Ex. (i) the values $t = \pm 1$ satisfy the equation $\frac{d^2y}{dt^2}\frac{dx}{dt} = \frac{d^2x}{dt^2}\frac{dy}{dt}$, but ρ does not $\to \infty$, for $\frac{dx}{dt}$, $\frac{dy}{dt}$ and $\frac{d^2y}{dt^2}\frac{dx}{dt} - \frac{d^2x}{dt^2}\frac{dy}{dt}$ all have a common factor $t^2 - 1$.

Or again,
$$x = t^3 + 2t,$$
$$y = 2t^3 + 3t^2 + t.$$

At O $\quad \dfrac{y}{x} \to \dfrac{1}{2}$ (for as $t \to 0$, $x \to 2t$, $y \to t$).

Hence the tangent is $\quad y = \dfrac{x}{2}.$

But $\qquad y - \dfrac{x}{2} = \dfrac{3t^3}{2} + 2t^2.$

Therefore the approximation is
$$y - \dfrac{x}{2} = 2t^3 = \dfrac{x^3}{2},$$

giving the branch $\quad y = \dfrac{x}{2} + \dfrac{x^3}{2}.$

EXAMPLES X B

Test for singularities the following curves:

(1) $x^3 + y^3 - x^2 - y^2 = 0.$ (2) $x^2 - 2xy + y^2 - x^3 = 0.$

(3) $y^2 - x^3 - x^2 = 0.$ (4) $x^{\frac{2}{3}} + y^{\frac{2}{3}} = a^{\frac{2}{3}}.$

(5) $y^2(a^2 + x^2) = x^2(a^2 - x^2).$ (6) $x = t^4 + 8t,\ y = t^3 + 1.$

(7) $x = t^2 + t + 1,\ y = t^3 + t^2 + t + 1.$

(8) $x = \dfrac{(t+1)^6}{t^4},\ \ y = \dfrac{(t+1)^4}{t^3}.$

(9) Show that the curve
$$x = 3t^4 - 4t^3 - 6t^2 + 12t,\ \ y = 8t^3 - 12t^2 + 9$$
has a cusp at $(5, 5)$.

(10) Show that the cycloid
$$x = a(\theta + \sin\theta),\ \ y = a(1 - \cos\theta)$$
has a succession of cusps with cuspidal tangents parallel to OY.

(11) Show that a first approximation to the shape of
$$x = t^2 + t,\ \ y = t^3 + t^2 + t$$
at the origin is $\qquad y - x = x^3.$

(12) Prove that the curve
$$\left.\begin{array}{l} x = t^4 - t^2 + 1 \\ y = t^3 - t^2 \end{array}\right\}$$
has a node at $(1, 0)$, but no real inflexions.

CHAPTER XI

RECTILINEAR ASYMPTOTES

§ 114. The Directions of the Curve at Infinity.

The directions of the branches of a curve which approach infinity are given by the factors of the highest homogeneous set.

This is easily seen if we put the equation into polar coordinates, for it must take the form

$$r^n f_n (\cos \theta, \sin \theta) + r^{n-1} f_{n-1} (\cos \theta, \sin \theta) + \ldots$$
$$+ \ldots = 0,$$

or, dividing by r^n,

$$f_n (\cos \theta, \sin \theta) + \frac{1}{r} f_{n-1} (\cos \theta, \sin \theta) + \ldots = 0.$$

If we put any particular value for θ in this last equation we have an equation in $\frac{1}{r}$, giving the reciprocals of the distances at which a line through O in this particular direction cuts the curve.

Now, if $f_n (\cos \theta, \sin \theta) = 0$, $\frac{1}{r}$ is a factor, and hence one value of the distance-reciprocal is zero; that is, the curve must have one point infinitely distant in each of the directions given by the equation $f_n (\cos \theta, \sin \theta) = 0$. If we convert this equation back to Cartesians by multiplying by r^n, we get $f_n (x, y) = 0$; that is, we equate to zero the highest homogeneous set of terms.

§ 115. Asymptotes.

An asymptote is a line to which a branch of a curve continually approaches in such a way that the branch is indistinguishable from the curve at infinity. The student is no doubt familiar with asymptotes in connection with the hyperbola; we also notice in passing the fact that, since any algebraic curve is continuous, a curve appears to have two branches to its asymptote, one approaching it, the other receding from it, but analytically these are considered as one branch. The factors of the highest homogeneous set do not in general give the actual asymptotes, but only lines through O parallel to them, and to get the actual asymptotes a closer approximation is required.

§ 116. The Reductions at Infinity.

It is clear that at infinity as well as at O we can safely use the homogeneous reduction, viz. that of putting $y = mx$ except in the factor $(y - mx)$, for this depends solely on the fact that the ratio $\frac{y}{x} \to m$, and this is true at infinity along $y = mx$ and along any line parallel to it, if $(y - mx)$ is a factor of the highest set.

But we cannot be safe in employing the division reduction, for this depends on a comparison of the terms $x^a \beta^b$ and $x^{a+a'} \beta^{b+b'}$. Now if we consider the various ways in which curves "go to infinity" we shall see that three distinct things may happen to β, which is, after all, an approximation to the divergence of the curve from the line $y = mx$.

(i) β may be zero ($y = mx$ itself would be the asymptote).

(ii) β may be finite (a line parallel to $y = mx$ is the asymptote).

(iii) β may be itself infinite. (The curve now tends to become parallel to $y = mx$, but at an infinite distance from it. This is called a "parabolic branch," for it represents the manner in which the parabola $y^2 = 4ax$ approaches its infinite direction $y = 0$.)

In each of these cases the residue $x^{a'}\beta^{b'}$, left over when one term is divided by another, has a different value, and the method is manifestly unsafe.

§ 117. Asymptotes parallel to the Axes.

We proceed to give methods for finding the actual asymptotes.

For an asymptote parallel to $x = 0$, equate to zero the coefficient of the highest power of y (whether containing x or not). If this leads to the equation of a straight line this straight line is an asymptote.

Similarly for an asymptote parallel to $y = 0$, equate to zero the highest power of x.

We can prove these results as follows. Write the equation to the curve in the form

$$a_0 y^n + a_1 y^{n-1}x + a_2 y^{n-2}x^2 + \ldots$$
$$+ b_1 y^{n-1} \quad + b_2 y^{n-2}x \quad + \ldots$$
$$+ c_2 y^{n-2} \quad + \ldots$$
$$+ \ldots = 0,$$

or, collecting terms,

$$a_0 y^n + y^{n-1}(a_1 x + b_1) + y^{n-2}(a_2 x^2 + b_2 x + c_2) + \ldots = 0.$$

Suppose that one of the infinite directions is $x = 0$; then $a_0 = 0$, for x must be a factor of the highest homogeneous set. Now the condition that an asymptote parallel to $x = 0$ exists is that, whilst x has some particular finite value, y becomes infinite, and further that *two* roots of the equation for y become infinite, for an asymptote, by its definition, is seen to partake of the general property of a tangent. But two roots of the equation in y will become infinite only if the two highest powers of y vanish. Therefore $a_0 = 0$ (as we saw already) and $a_1 x + b_1 = 0$.

Hence this value of x, obtained by equating to zero the coefficient of the highest surviving power of y, gives the equation to the asymptote. If this term happens to be absent in the equation, the next lowest term must be equated to zero, giving two parallel asymptotes, and so on. Similarly the rule for asymptotes parallel to $y = 0$ can be proved.

§ 118. Oblique Asymptotes.

The method for oblique directions is the same, but we must first use the homogeneous reduction. The reduced equation becomes

$$(a_1 \beta + b_1) x^{n-1} + (a_2 \beta^2 + b_2 \beta + c_2) x^{n-2} + \ldots = 0.$$

(N.B. No term $a_0 x^n$ exists, since by hypothesis $(y - mx)$, i.e. β, is a factor of the highest set.)

If, in addition, $a_1 \beta + b_1 = 0$, another value of x becomes infinite and $y = mx + \beta$,

i.e. $$y = mx - \frac{b_1}{a_1}$$

is a rectilinear asymptote.

If, by any chance, the work of this article or the previous one leads to an equation of the form Constant $= 0$, then no finite value of β exists, and the curve must " go to infinity " in a direction parallel to $y = mx$ but at a very great distance from it. Cf. § 116 (iii).

§ 119. Equations of form $F_n + F_{n-2} + \ldots = 0$.

It can easily be proved that if no homogeneous set of order $(n-1)$ exists, the highest set gives the actual asymptotes, provided the highest set contains no repeated factor.

For suppose $y = mx$ is any factor of the highest set, and let the homogeneous reduction be performed. We now get a reduced equation

$$a_0 x^n + a_1 x^{n-1}\beta + a_2 x^{n-2}\beta^2 + \ldots$$
$$+ c_2 x^{n-2} + \ldots$$
$$+ \ldots = 0.$$

As before, a_0 necessarily $= 0$ and for another value of x to become infinite we need $a_1\beta = 0$, i.e. $\beta = 0$.

Therefore $y = mx$ is the actual asymptote, and similarly for any of the other infinite directions.

Thus, the asymptotes of $F_n + F_{n-2} + \ldots = 0$ are given by $F_n = 0$, a theorem which is often of use. The way in which this breaks down when there is a repeated factor is shown in § 120, Ex. (iii).

The knowledge of the form at O and the equations to the asymptotes, together with obvious considerations such as the intersections of the curve with the axes and so on, will often enable a curve to be traced completely, as we shall see in the third example of the following section.

§ 120. Worked Examples.

(i) *Find the asymptotes of*
$$(x + y)^3 (x - y)^2 - 2a (x + y)^2 (x - y)^2$$
$$- 2a^2 (x^3 + y^3) (x + y) + 2a^3 (x - y)^2 + 4a^4 (x - y) = 0.$$

The directions at infinity are $x + y = 0$ and $x - y = 0$.

For the direction $y + x = 0$.

Put $y = -x$, but $y + x = \beta$.

$\therefore\quad 4\beta^3 x^2 - 8a\beta^2 x^2 - 4a^2\beta x^2 + 8a^3 x^2 + 8a^4 x = 0$.

Equate to zero the coefficient of x^2.

$\therefore\quad 4\beta^3 - 8a\beta^2 - 4a^2\beta + 8a^3 = 0$,

and hence $\beta = -2a$, a, or $-a$.

We thus have the three parallel asymptotes

$$\left.\begin{array}{l} y + x = -2a \\ y + x = \quad a \\ y + x = - \quad a \end{array}\right\}.$$

For the direction $y = x$.

Put $y = x$, but $y - x = \beta$.

$\therefore\quad 8\beta^2 x^3 - 8a\beta^2 x^2 - 8a^2 x^3 + 2a^3\beta^2 - 4a^4\beta = 0$.

Therefore, equating to zero the coefficient of x^3, we have

$$8(\beta^2 - a^2) = 0 \quad\text{or}\quad \beta = \pm a.$$

This direction gives two parallel asymptotes

$$\left.\begin{array}{l} y = x + a \\ y = x - a \end{array}\right\}.$$

(ii) *Find the asymptotes of* $x^3(y^2 - a^2) = y^3(x^2 - 8a^2)$.

The equation rearranged in its homogeneous sets is

$$x^2 y^2 (x - y) - a^2 (x^3 - 8y^3) = 0.$$

The directions at infinity are $x = 0$, $y = 0$, $x = y$.

For $x = 0$. We equate to zero the coefficient of y^3.

This gives $\qquad x^3 - 8a^2 = 0$,

i.e. there are two asymptotes $x = \pm 2\sqrt{2}a$.

For $y = 0$. We equate to zero the coefficient of x^3, i.e.

$$y^2 - a^2 = 0.$$

This gives two asymptotes $y = \pm a$.

For $y = x$. Put $y = x$, but $y - x = \beta$.

$$\therefore \quad x^4 \beta + 7a^2 x^3 = 0.$$

Equate to zero the coefficient of x^4; then $\beta = 0$.

Hence $y = x$ itself is an asymptote.

Now we can derive a little further information.

For, from the reduced equation $\beta = - \dfrac{7a^2}{x}$, which is

positive when x is negative and negative when x is positive. Thus the curve lies below its asymptote. This step can however only be done if we have previously found $\beta = 0$.

(iii) *Trace the curve* $(x^2 - y^2)^2 + 4a^2 xy = 0$.

The directions at O are $x = 0$, $y = 0$.

Put $y = 0$, except in the factor y; we get

$$x^3 + 4a^2 y = 0, \text{ giving branch } A.$$

Put $x = 0$, except in the factor x; we get

$$y^3 + 4a^2 x = 0, \text{ giving branch } B.$$

At infinity. The directions at infinity are $y = \pm x$.

Put $y = x$, but $y - x = \beta$.

$$\therefore \quad 4\beta^2 x^2 + 4a^2 x^2 = 0.$$

Equate to zero the coefficient of x^2.

$$\therefore \quad \beta^2 + 4a^2 = 0.$$

Therefore β is imaginary.

Hence there are no real asymptotes in this direction.

Put $y = -x$, but $y + x = \beta$.

$$\therefore \quad 4\beta^2 x^2 - 4a^2 x^2 = 0.$$

Equate to zero the coefficient of x^2.

$$\therefore \quad \beta = \pm a.$$

We therefore get two parallel asymptotes

$$y + x = \pm a.$$

(Note that the theorem of § 119 does not hold for this curve, as there are repeated factors in the highest homogeneous set.)

But, by putting in succession $x = 0$, $y = 0$ and $y = \pm x$, we can see that the axes and the lines $y = \pm x$ are not crossed except at O.

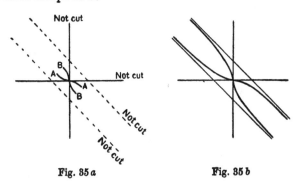

Fig. 35 a Fig. 35 b

Also the curve cannot cross either of its asymptotes, for if we put $y + x = a$ we get

$$a^2 (2x - a)^2 + 4a^2 x (a - x) = 0,$$

i.e. $$a^4 = 0.$$

Hence our rough scheme for the curve is as in Fig. 35 a; and the curve is as in Fig. 35 b.

§ 121. Parallel Asymptotes. Special Considerations.

The above considerations are sufficient to enable many simple curves to be traced. Special caution, however, must be observed in regard to curves having parallel asymptotes.

Consider a finite node of a curve, as in Fig. 36 *a*. Each tangent at the node meets the curve in three (ultimately coincident) points, two of them belonging to its own branch, and one to the other branch.

Suppose now that the node is removed to a point at infinity. The two tangents become tangents having their points of contact at infinity, i.e. asymptotes; and further, since they meet at the same point at infinity, they must be parallel asymptotes. Evidently the nature of the figure is unaltered, and hence the lines will still intersect the curve at three points, but these points are now at infinity. It follows that in counting up the intersections of the asym-

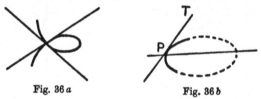

Fig. 36 *a*   Fig. 36 *b*

ptotes with the curve (a course which is often necessary to verify our drawing) we must count each of a pair of parallel asymptotes as meeting the curve at three points at infinity, and not at two, as we normally expect an asymptote to do.

The force of this statement can be appreciated by reference to Ex. (iii) of § 120. Now suppose the asymptote $y + x = a$ had been proved, by an error in the algebra, to cut the curve at some finite point. Since $y = x$ is an axis of symmetry (see § 101 (3)), the asymptote would also have to cut the curve in another point, the reflexion of the first. The intersections of the asymptote with the curve would now be five in number, viz. the two finite intersections, two at infinity on the branch belonging to the asymptote itself, and another also at infinity on the branch belonging to the other asymptote. But it is manifestly impossible for a quartic curve to be cut by a straight line in five points; and hence our hypothesis of a finite intersection with one of the asymptotes must break down.

Considerations of this character will often save a mistake, or aid in detecting one that has been made.

§ 122. Contact at Infinity.

One further consideration is of interest.

Consider an ordinary tangent to a curve at a finite point P. If the contact is of the first order the curve does not cross its tangent, and the general shape of the curve near the point of contact is elliptical (as in Fig. 36 b). Now consider any other line through P, and let this line be projected to infinity. The ellipse which denotes the approximate shape of the curve at P becomes a conic cutting the line at infinity in real points, i.e. a hyperbola; and the tangent becomes an asymptote. But we know that the two branches of a hyperbola are on opposite sides of the asymptotes. Hence the normal behaviour of a branch having ordinary contact with its asymptote is to return on the other side of it. Similarly a curve which returns on the *same* side of its asymptote is the projection of an inflexion, and thus the asymptote is cut in three points at infinity (at least).

We can employ this consideration also to verify the curve given in § 120 (iii).

For conditions of symmetry cause it to return on the same side of its asymptote, thus accounting for three intersections; and, as in § 121, the parallel asymptote accounts for another. Thus four intersections of the curve with either asymptote are accounted for, and these are all that can exist.

Space prohibits any further discussion of curves at infinity, particularly in relation to parabolic branches and curvilinear asymptotes. For further information the student is referred to Hilton's *Plane Algebraic Curves* and Frost and Bell's *Curve Tracing*.

§ 123. Polar Asymptotes.

It is not often necessary to find asymptotes in polar coordinates, this system of coordinates being in general more adapted to the expression of properties of closed and recurring curves. If it is necessary to find them, however, the best method is to find the equation of the tangent at the point where $\theta = \alpha$, and then to determine the limiting form of this equation for those values of α for which $r \to \infty$.

The figure should always be drawn for each case and due regard paid to signs of expressions involved.

EXAMPLES XI

Find the asymptotes of :

(1) $(x^2 + y^2)(3x + 4y) = 3x^2 - 10xy + 3y^2$.

(2) $y(y-x)^2(y-2x) + 3x^2(y-x) - 2x^2 = 0$.

(3) $y^2(y-x)^2 - 3x^2 + 4x + y = 0$.

(4) $x^4 - y^4 + 3x^2y + 3xy^2 + xy = 0$.

(5) $x^3(y^2 - x^2) + xy^2 + x^2y + y = 0$.

(6) $x^3(y^2 - x^2) + xy^2 + x^2y + y^2 = 0$.

(7) $x^3(y^2 - x^2) + xy(x+y) + 2y^2 = 0$.

(8) $x(x^2 - y^2) = 4x^2 - 6xy - 4y^2$.

(9) $x^2y + 2xy^2 + 6xy + 2x^3 + 4(x-y) = 0$.

(10) $x^3(x-y)^2 - 2y(x-y) + y = 0$.

(11) $2x(x-y)^2 - 3a(x^2 - y^2) + 4a^2y = 0$.

(12) $y^4 - x^4 = xy$.   (13) $x^3(x+y) = ay^2$.

(14) $xy(x^2 + y^2) = x^2 - y^2$.   (15) $xy(x^2 - y^2) = x^2 + y^2$.

(16) $x^2y^2 = (x+a)^2(x^2 + a^2)$.

Trace the following curves:

(17) $x^3 + y^3 = 3axy$.

(18) $x^5 + y^5 - 5x^2y^2 = 0$.

(19) $(x^2 + y^2)(3x + 4y) = 3x^2 - 10xy + 3y^2$.

(20) $(x^2 - a^2)(x - a)^2 + (y^2 - a^2)^2 = 0$.

(21) $(x - y)^2(x + 2y) + (x + y) = 0$.

(22) $(x - y)^2(x + 2y) - (x + y) = 0$.

(23) $y(x - y)^2 = 2(x + y)$.

(24) $x^3 + y^3 = (y - x)(y - 2x)$.

(25) $y^2(x^2 - y^2) - 2y^3 + 2x = 0$.

(26) $y^2(x - 1)(x - 2) - x^2 = 0$.

(27) $y(x - y)^2 - x = 0$.

(28) $y^4 - x^4 = a^2xy$.

(29) $(y - x)(y - x^3) = x^4y^2$.

*(30) Prove that all the finite intersections of a curve with its asymptotes lie on a curve of degree $n - 2$.

*(31) A cubic curve has $x + a = 0$ and $y + b = 0$ as asymptotes and a node at O. Prove that the finite intersection of the curve with its third asymptote must lie on $bx + ay = 0$.

By considering values of r as θ varies between 0 and 2π, trace the following polar curves:

(32) $r = a \sin 4\theta$.

(33) $r = a \sin 5\theta$.

(34) $r^3 = a^2 \cos 2\theta$.

(35) $r = \dfrac{a\theta}{\theta + 1}$.

(36) Show that all curves of the form $r = a \sin n\theta$ fall into two groups, having n loops if n is odd, but $2n$ loops if n is even.

(37) Show also that all curves of the form $r = a \sin n\theta$ lie within a certain circle.

(38) Find the asymptotes of $r \sin 4\theta = a$, and trace the curve.

(39) Find the asymptote of $r = \dfrac{a\theta}{\theta + 1}$.

CHAPTER XII

PARTIAL DIFFERENTIATION

§ 124. Nature of Partial Differentiation.

We have already (Ch. III, § 38) called attention to the notation $\dfrac{\partial u}{\partial x}, \dfrac{\partial u}{\partial y}$, which is employed for the partial derivatives of a function; and we have quoted, without any proof other than by trial, the formula for differentiating an implicit function, viz.

$$\frac{dy}{dx} = -\frac{\dfrac{\partial u}{\partial x}}{\dfrac{\partial u}{\partial y}}.$$

We collect here a few of the more important results obtained for these partial differential coefficients. It should be clearly understood from the outset, that these differential coefficients are subject to more restrictions than ordinary differential coefficients, for, by their nature, they impose restrictions on the variables.

We cannot, for instance, say safely that

$$\frac{\partial y}{\partial x}\frac{\partial x}{\partial y} = 1,$$

or that

$$\frac{\partial y}{\partial z} = \frac{\partial y}{\partial x}\frac{\partial x}{\partial z}.$$

For each of these differential coefficients is calculated under a different hypothesis as to the variables; and we shall, in fact, see that the latter of these two statements is incorrect.

§ 125. The Theorem of the Total Differential.

The important result on which most of the theory of partial differentiation is based is known as the Theorem of the Total Differential. We proceed to give this result for three variables, and it will be seen that the principle is equally applicable to any number of variables.

Let $\phi(x, y, z)$ be some function of the variables x, y, z, and let t be a parameter in terms of which x, y, z can all be expressed. Then the theorem is that

$$\frac{d\phi}{dt} = \frac{\partial\phi}{\partial x}\frac{dx}{dt} + \frac{\partial\phi}{\partial y}\frac{dy}{dt} + \frac{\partial\phi}{\partial z}\frac{dz}{dt},$$

provided that $\qquad \phi, \quad \dfrac{\partial\phi}{\partial x}, \quad \dfrac{\partial\phi}{\partial y}, \quad \dfrac{\partial\phi}{\partial z}$

are all continuous; or, as we often write it, in order to avoid throwing any undue emphasis on the parameter t,

$$d\phi = \frac{\partial\phi}{\partial x}\,dx + \frac{\partial\phi}{\partial y}\,dy + \frac{\partial\phi}{\partial z}\,dz *.$$

Now suppose that as t increases to $t + t'$, x increases as a result to $x + x'$, y to $y + y'$, and z to $z + z'$.

$$\therefore \ \phi(x, y, z) + \delta\phi(x, y, z) = \phi(x + x', y + y', z + z').$$

$$\therefore \ \delta\phi = \phi(x + x', y + y', z + z') - \phi(x, y, z)$$

$$= x'\left[\frac{\phi(x + x', y + y', z + z') - \phi(x, y + y', z + z')}{x'}\right]$$

$$+ y'\left[\frac{\phi(x, y + y', z + z') - \phi(x, y, z + z')}{y'}\right]$$

$$+ z'\left[\frac{\phi(x, y, z + z') - \phi(x, y, z)}{z'}\right],$$

as may easily be verified by simplifying the three fractions.

---

* As in footnote, p. 25, we may also say

$$\delta\phi = \frac{\partial\phi}{\partial x}\cdot\delta x + \frac{\partial\phi}{\partial y}\cdot\delta y + \frac{\partial\phi}{\partial z}\cdot\delta z \text{ approximately,}$$

where $\delta\phi$, δx, δy, δz are actual changes in ϕ, x, y, z.

$$\therefore \operatorname*{Lt}_{t' \to 0} \frac{\delta\phi}{t'}$$

$$= \operatorname*{Lt}_{t' \to 0} \frac{x'}{t'} \left[\frac{\phi(x + x', y + y', z + z') - \phi(x, y + y', z + z')}{x'} \right]$$

$$+ \operatorname*{Lt}_{t' \to 0} \frac{y'}{t'} \left[\frac{\phi(x, y + y', z + z') - \phi(x, y, z + z')}{y'} \right]$$

$$+ \operatorname*{Lt}_{t' \to 0} \frac{z'}{t'} \left[\frac{\phi(x, y, z + z') - \phi(x, y, z)}{z'} \right].$$

Now clearly the limit of $\frac{\delta\phi}{t'}$ is $\frac{d\phi}{dt}$, and similarly the limits of $\frac{x'}{t'}, \frac{y'}{t'}, \frac{z'}{t'}$ are $\frac{dx}{dt}, \frac{dy}{dt}, \frac{dz}{dt}$ respectively.

Now as $t' \to 0$, $x' \to 0$ also. Hence

$$\operatorname*{Lt}_{t' \to 0} \left[\frac{\phi(x + x', y + y', z + z') - \phi(x, y + y', z + z')}{x'} \right]$$

$$= \operatorname*{Lt}_{x' \to 0} \left[\frac{\phi(x + x', y + y', z + z') - \phi(x, y + y', z + z')}{x'} \right]$$

$$= \operatorname*{Lt}_{x' \to 0} \left[\frac{\phi(x + x', y + y', z + z') - \phi(x, y + y', z + z')}{x + x' - x} \right]$$

$$= \operatorname*{Lt}_{x' \to 0} \left[\phi'(x + \theta x', y + y', z + z') \right],$$

using the First Theorem of the Mean (§ 70), and assuming, as usual, that we are dealing with continuous functions, at any rate, as far as x is concerned.

But since $x + \theta x' \to x$ as $x' \to 0$, we get

$$\phi'(x, y + y', z + z'),$$

i.e. a differentiation performed whilst keeping y and z unchanged, and this is rightly defined by $\frac{\partial\phi}{\partial x}$.

By a similar argument we see that

$$\underset{t' \to 0}{\mathrm{Lt}} \left[\frac{\phi(x, y + y', z + z') - \phi(x, y, z + z')}{y'} \right]$$

leads us to
$$\frac{\partial \phi}{\partial y},$$

and similarly

$$\underset{t' \to 0}{\mathrm{Lt}} \left[\frac{\phi(x, y, z + z') - \phi(x, y, z)}{z'} \right]$$

is
$$\frac{\partial \phi}{\partial z}.$$

Hence the theorem becomes

$$\frac{d\phi}{dt} = \frac{\partial \phi}{\partial x}\frac{dx}{dt} + \frac{\partial \phi}{\partial y}\frac{dy}{dt} + \frac{\partial \phi}{\partial z}\frac{dz}{dt}.$$

We should also note that any variable can be chosen as t, even one of the original variables. For instance, if we put x in place of t, we get

$$\frac{d\phi}{dx} = \frac{\partial \phi}{\partial x} + \frac{\partial \phi}{\partial y}\frac{dy}{dx} + \frac{\partial \phi}{\partial z}\frac{dz}{dx}.$$

§ 126. The Total Differential (*continued*).

The difference between $\dfrac{\partial \phi}{\partial x}$ and $\dfrac{d\phi}{dx}$ should be carefully noticed. The second of these expressions is called a " total " differential to distinguish it from a " partial " differential. The first is the limiting value of the average increase dependent on x only; but the second also takes into account the corresponding changes in the value of y and z produced by the change in the variable x, these being denoted by the last two terms in the equation giving $\dfrac{d\phi}{dx}$.

We can also see that if x, y, z do not depend totally

upon one parameter t, but partly on a parameter t and partly on another parameter s, then

$$\frac{\partial \phi}{\partial t} = \frac{\partial \phi}{\partial x}\frac{\partial x}{\partial t} + \frac{\partial \phi}{\partial y}\frac{\partial y}{\partial t} + \frac{\partial \phi}{\partial z}\frac{\partial z}{\partial t}.$$

For there is nothing in the argument to preclude the existence of another parameter, so long as it is not allowed to change its value during the process of extracting the limits, and this is precisely what is implied by $\frac{\partial \phi}{\partial t}$.

The latter form of the Total Differential is also sometimes written in the form

$$\partial \phi = \frac{\partial \phi}{\partial x}\,\partial x + \frac{\partial \phi}{\partial y}\,\partial y + \frac{\partial \phi}{\partial z}\,\partial z.$$

§ 127. Differentiation of Implicit Functions.

We can now give a proof of the rule given in Ch. III for the differentiation of an implicit function.

For let $u(x, y) = 0$ be an implicit relation connecting x and y; and suppose that x and y can both be expressed in terms of a parameter t.

Then, by the Theorem of the Total Differential,

$$\frac{du}{dt} = \frac{\partial u}{\partial x}\frac{dx}{dt} + \frac{\partial u}{\partial y}\frac{dy}{dt}.$$

But, since $u(x, y) = 0$, $\frac{du}{dt} = 0$ (for the relation $u = 0$ is satisfied for all values given to t),

$$\therefore \frac{\partial u}{\partial x}\frac{dx}{dt} = -\frac{\partial u}{\partial y}\frac{dy}{dt}.$$

$$\therefore \frac{dy}{dx} = \frac{\dfrac{dy}{dt}}{\dfrac{dx}{dt}} = -\frac{\dfrac{\partial u}{\partial x}}{\dfrac{\partial u}{\partial y}}.$$

In actual practice we never trouble about the existence of the parameter t. Though it clearly must exist, it is often a very difficult matter to determine what it is.

§ 128. Partial Differentials.

The same theorem can be employed to find partial derivatives in cases where these would be difficult to calculate directly. This is shown by the following example.

Ex. If $x^3 + y^3 + z^3 - 3xyz = 0$, find $\dfrac{\partial z}{\partial x}$ and $\dfrac{\partial z}{\partial y}$.

Now $\dfrac{\partial z}{\partial x}$ and $\dfrac{\partial z}{\partial y}$ imply that z is to be regarded as a function of x and y, i.e. that we have to bring the given equation into the form $z = \phi(x, y)$; but, since the equation is a cubic in z, this would be an exceedingly laborious piece of work.

To avoid this, we proceed as follows.

Let $\qquad u \equiv x^3 + y^3 + z^3 - 3xyz = 0,$

$$\therefore \frac{du}{dt} \equiv \frac{\partial u}{\partial x}\frac{dx}{dt} + \frac{\partial u}{\partial y}\frac{dy}{dt} + \frac{\partial u}{\partial z}\frac{dz}{dt}$$

$$= 3(x^2 - yz)\frac{dx}{dt} + 3(y^2 - zx)\frac{dy}{dt} + 3(z^2 - xy)\frac{dz}{dt}$$

$$= 0.$$

But, in calculating $\dfrac{\partial z}{\partial x}$, we have to keep y constant, i.e. we must make $\dfrac{dy}{dt} = 0$.

Hence $\qquad 3(x^2 - yz)\dfrac{dx}{dt} + 3(z^2 - xy)\dfrac{dz}{dt} = 0.$

$$\therefore \left(\frac{dz}{dx}\right)_{y\text{ const.}} = -\frac{x^2 - yz}{z^2 - xy},$$

i.e. $\qquad\qquad \dfrac{\partial z}{\partial x} = -\dfrac{x^2 - yz}{z^2 - xy}.$

Likewise, if we wish to get $\dfrac{\partial z}{\partial y}$, we put $\dfrac{dx}{dt} = 0$ and so on.

Proceeding in this way we get all the partial derivatives, and the reader will easily verify that

$$\frac{\partial z}{\partial x}\frac{\partial x}{\partial y}\frac{\partial y}{\partial z} = -1$$

in this case, and not $+1$, as one might suppose at first thought.

The relations between the Cartesian and polar coordinates of a point provide a useful basis for examples of this sort. For instance, we may assign two meanings to the expression $\dfrac{\partial x}{\partial r}$, and it is easy to show that the product of these two values is unity.

For there are two equations which contain both x and r, viz.

$$r = \sqrt{x^2 + y^2}$$

or

$$x = \sqrt{r^2 - y^2}$$

and

$$x = r \cos \theta.$$

From the first of these we could legitimately get a value of $\dfrac{\partial x}{\partial r}$ by keeping y constant, and from the second a value of $\dfrac{\partial x}{\partial r}$ by keeping θ constant.

We should get

$$\left(\frac{\partial x}{\partial r}\right)_{y \text{ const.}} = \frac{r}{\sqrt{r^2 - y^2}}$$

and

$$\left(\frac{\partial x}{\partial r}\right)_{\theta \text{ const.}} = \cos \theta.$$

But

$$\frac{r}{\sqrt{r^2 - y^2}} = \frac{r}{\sqrt{r^2 - r^2 \sin^2 \theta}} = \frac{1}{\cos \theta}.$$

$$\therefore \left(\frac{\partial x}{\partial r}\right)_{y \text{ const.}} \times \left(\frac{\partial x}{\partial r}\right)_{\theta \text{ const.}} = 1.$$

As a matter of fact, if $\frac{\partial x}{\partial r}$ is written without any restriction, it should be taken to mean the second of these, which is obtained from the equation in which x is given in terms of the two polar coordinates: and similarly, if we have $\frac{\partial r}{\partial x}$, we derive it from that equation by which r is given in terms of the two Cartesian coordinates.

§ 129. Euler's Theorem on Homogeneous Functions.

The Total Differential affords an interesting proof of Euler's Theorem.

If f (x, y, z) is a homogeneous function of degree n in x, y, z, then

$$x \frac{\partial f}{\partial x} + y \frac{\partial f}{\partial y} + z \frac{\partial f}{\partial z} = nf (x, y, z).$$

We can formulate the general definition of a homogeneous function as follows.

If in any function tx, ty and tz are put instead of x, y, z and the result is $t^n \times f(x, y, z)$, then the function is called a *homogeneous function of degree n*. This definition includes functions that are not algebraic, though, of course, many well-known homogeneous functions are algebraic. For example, $x^3 + y^3 + z^3 - 3xyz$ is a homogeneous function of degree 3, for if we put tx, etc. for x, y, z, we get

$$t^3 (x^3 + y^3 + z^3 - 3xyz),$$

but
$$\sin^{-1} \frac{x + y + z}{\sqrt{x^2 + y^2 + z^2}},$$

though not algebraic, is a homogeneous function of degree 0.

Euler's Theorem for these two cases would give

(i) $x \dfrac{\partial f}{\partial x} + y \dfrac{\partial f}{\partial y} + z \dfrac{\partial f}{\partial z} = 3 (x^3 + y^3 + z^3 - 3xyz),$

(ii) $x \dfrac{\partial f}{\partial x} + y \dfrac{\partial f}{\partial y} + z \dfrac{\partial f}{\partial z} = 0,$

both of which results can be verified by elementary methods.

The proof is as follows.

Let $u \equiv f(x, y, z)$ be a homogeneous function of degree n in three variables, and let x_0, y_0, z_0 be a set of values which remain unchanged throughout the operation, the corresponding value of the function being u_0.

If we now put $x = tx_0$, $y = ty_0$, $z = tz_0$, we get, since the function is homogeneous,

$$u = t^n u_0.$$

$$\therefore \frac{du}{dt} = n t^{n-1} u_0.$$

$$\therefore t \frac{du}{dt} = n t^n u_0 = nu.$$

But, by the Total Differential,

$$\frac{du}{dt} = \frac{\partial u}{\partial x} \frac{dx}{dt} + \frac{\partial u}{\partial y} \frac{dy}{dt} + \frac{\partial u}{\partial z} \frac{dz}{dt}$$

$$= x_0 \frac{\partial u}{\partial x} + y_0 \frac{\partial u}{\partial y} + z_0 \frac{\partial u}{\partial z}. \quad \text{(For } x = x_0 t, \text{ etc.)}$$

$$\therefore t \frac{du}{dt} = tx_0 \frac{\partial u}{\partial x} + ty_0 \frac{\partial u}{\partial y} + tz_0 \frac{\partial u}{\partial z}$$

$$= x \frac{\partial u}{\partial x} + y \frac{\partial u}{\partial y} + z \frac{\partial u}{\partial z}.$$

Hence $\qquad x \dfrac{\partial u}{\partial x} + y \dfrac{\partial u}{\partial y} + z \dfrac{\partial u}{\partial z} = nu.$

The proof is seen to be equally valid for any number of variables.

§ 130. Partial Derivatives of Higher Order.

The expressions $\frac{\partial^2 u}{\partial x^2}$, $\frac{\partial^2 u}{\partial y^2}$, $\frac{\partial^2 u}{\partial x \partial y}$ are called partial derivatives of the second order. The meaning of the first is clear. It is that we first find the partial derivative $\frac{\partial u}{\partial x}$; this will be a function of (x, y), which can be differentiated again partially. And so for $\frac{\partial^2 u}{\partial y^2}$.

Thus, if $$u \equiv x^3 y^2 + x^2 y^3,$$

then $$\frac{\partial u}{\partial x} = 3x^2 y^2 + 2xy^3,$$

and $$\frac{\partial^2 u}{\partial x^2} = 6xy^2 + 2y^3.$$

So also $$\frac{\partial u}{\partial y} = 2x^3 y + 3x^2 y^2$$

and $$\frac{\partial^2 u}{\partial y^2} = 2x^3 + 6x^2 y.$$

With regard to the expression $\frac{\partial^2 u}{\partial x \partial y}$, it appears at first sight that some confusion may arise as to the order in which the differentiations are performed; i.e. whether $\frac{\partial^2 u}{\partial x \partial y}$ means $\frac{\partial}{\partial x} \left\{ \frac{\partial u}{\partial y} \right\}$ or $\frac{\partial}{\partial y} \left\{ \frac{\partial u}{\partial x} \right\}$. As a matter of fact, it is quite immaterial which of the two is adopted, since it has been proved that they must lead to the same result. But the proof involves difficult ideas on the subject of Limits and is beyond the scope of this book. The reader will find the question fully discussed in Goursat's *Cours d'Analyse*, Vol. I, Ch. I, § 11.

In the example used above, we see that

$$\frac{\partial}{\partial y}\left\{\frac{\partial u}{\partial x}\right\} = \frac{\partial}{\partial y}\left\{3x^2y^2 + 2xy^3\right\}$$

$$= 6x^2y + 6xy^2$$

and

$$\frac{\partial}{\partial x}\left\{\frac{\partial u}{\partial y}\right\} = \frac{\partial}{\partial x}\left\{3x^2y^2 + 2x^3y\right\}$$

$$= 6xy^2 + 6x^2y,$$

the theorem being thus verified for this function.

The partial derivatives of higher order are similarly defined, and in the same way the order of performing the differentiations makes no difference.

§ 131. Second Differential of an Implicit Function.

The partial derivatives $\dfrac{\partial u}{\partial x}, \dfrac{\partial u}{\partial y}, \dfrac{\partial^2 u}{\partial x^2}, \dfrac{\partial^2 u}{\partial x \partial y}$ and $\dfrac{\partial^2 u}{\partial y^2}$ are often denoted by the letters p, q, r, s, t.   Hence

$$r = \frac{\partial^2 u}{\partial x^2} = \frac{\partial p}{\partial x}$$

and

$$t = \frac{\partial q}{\partial y}.$$

Also

$$s = \frac{\partial p}{\partial y} \text{ or } \frac{\partial q}{\partial x}.$$

We shall employ this notation in what follows.

When x, y are connected by an implicit relation we have

$$\frac{dy}{dx} = -\frac{\dfrac{\partial u}{\partial x}}{\dfrac{\partial u}{\partial y}} = -\frac{p}{q}.$$

$$\therefore \frac{d^2y}{dx^2} = \frac{d}{dx}\left\{\frac{dy}{dx}\right\}$$

$$= \frac{d}{dx}\left\{-\frac{p}{q}\right\}$$

$$= \frac{p\dfrac{dq}{dx} - q\dfrac{dp}{dx}}{q^2}.$$

Now p is a function of x and y, and hence, by the Total Differential, we have

$$\frac{dp}{dx} = \frac{\partial p}{\partial x}\frac{dx}{dx} + \frac{\partial p}{\partial y}\frac{dy}{dx}$$

$$= r - \frac{p}{q}s$$

$$= \frac{qr - ps}{q}.$$

So also

$$\frac{dq}{dx} = \frac{\partial q}{\partial x} + \frac{\partial q}{\partial y}\frac{dy}{dx}$$

$$= s - \frac{p}{q}t$$

$$= \frac{qs - pt}{q}.$$

Hence we have

$$\frac{d^2y}{dx^2} = \frac{p\left\{\dfrac{qs - pt}{q}\right\} - q\left\{\dfrac{qr - ps}{q}\right\}}{q^2}$$

$$= -\frac{q^2r - 2pqs + p^2t}{q^3}.$$

This formula enables us to calculate $\dfrac{d^2y}{dx^2}$, which is often required in curvature problems, from an implicit function.

§ 132. An Important Transformation.

As an example of the use of partial differentials we give here a transformation which is frequently required in Higher Mechanics, viz. to transform the expression

$$\frac{\partial^2 u}{\partial x^2} + \frac{\partial^2 u}{\partial y^2}$$

into polar coordinates.

We leave as an exercise for the reader the verifying of the two results in Ex. XII, No. 8, viz.

$$\frac{\partial u}{\partial x} = \cos \theta \, \frac{\partial u}{\partial r} - \frac{\sin \theta}{r} \frac{\partial u}{\partial \theta},$$

$$\frac{\partial u}{\partial y} = \sin \theta \, \frac{\partial u}{\partial r} + \frac{\cos \theta}{r} \frac{\partial u}{\partial \theta},$$

these results applying to all functions of x and y.

Now
$$\frac{\partial^2 u}{\partial x^2} = \frac{\partial}{\partial x} \left\{ \frac{\partial u}{\partial x} \right\}$$

$$= \cos \theta \, \frac{\partial}{\partial r} \left\{ \frac{\partial u}{\partial x} \right\} - \frac{\sin \theta}{r} \frac{\partial}{\partial \theta} \left\{ \frac{\partial u}{\partial x} \right\}$$

(for $\dfrac{\partial u}{\partial x}$ is a function of x, y and transforms in exactly the same way as u)

$$= \cos \theta \, \frac{\partial}{\partial r} \left\{ \cos \theta \, \frac{\partial u}{\partial r} - \frac{\sin \theta}{r} \frac{\partial u}{\partial \theta} \right\}$$

$$\qquad - \frac{\sin \theta}{r} \frac{\partial}{\partial \theta} \left\{ \cos \theta \, \frac{\partial u}{\partial r} - \frac{\sin \theta}{r} \frac{\partial u}{\partial \theta} \right\}$$

$$= \cos \theta \left\{ \cos \theta \, \frac{\partial^2 u}{\partial r^2} + \frac{\sin \theta}{r^2} \frac{\partial u}{\partial \theta} - \frac{\sin \theta}{r} \frac{\partial^2 u}{\partial r \partial \theta} \right\}$$

$$- \frac{\sin \theta}{r} \left\{ -\sin \theta \, \frac{\partial u}{\partial r} + \cos \theta \, \frac{\partial^2 u}{\partial r \partial \theta} - \frac{\cos \theta}{r} \frac{\partial u}{\partial \theta} - \frac{\sin \theta}{r} \frac{\partial^2 u}{\partial \theta^2} \right\}$$

$$= \cos^2 \theta \, \frac{\partial^2 u}{\partial r^2} - \frac{2 \sin \theta \cos \theta}{r} \frac{\partial^2 u}{\partial r \partial \theta} + \frac{\sin^2 \theta}{r^2} \frac{\partial^2 u}{\partial \theta^2}$$

$$\qquad + \frac{2 \cos \theta \sin \theta}{r^2} \frac{\partial u}{\partial \theta} + \frac{\sin^2 \theta}{r} \frac{\partial u}{\partial r}.$$

Similarly we have

$$\frac{\partial^2 u}{\partial y^2} = \sin\theta \frac{\partial}{\partial r}\left\{\frac{\partial u}{\partial y}\right\} + \frac{\cos\theta}{r}\frac{\partial}{\partial\theta}\left\{\frac{\partial u}{\partial y}\right\}$$

$$= \sin\theta \frac{\partial}{\partial r}\left\{\sin\theta \frac{\partial u}{\partial r} + \frac{\cos\theta}{r}\frac{\partial u}{\partial\theta}\right\}$$

$$+ \frac{\cos\theta}{r}\frac{\partial}{\partial\theta}\left\{\sin\theta \frac{\partial u}{\partial r} + \frac{\cos\theta}{r}\frac{\partial u}{\partial\theta}\right\}$$

$$= \sin^2\theta \frac{\partial^2 u}{\partial r^2} + \frac{2\sin\theta\cos\theta}{r}\frac{\partial^2 u}{\partial r\partial\theta} + \frac{\cos^2\theta}{r^2}\frac{\partial^2 u}{\partial\theta^2}$$

$$+ \frac{\cos^2\theta}{r}\frac{\partial u}{\partial r} - \frac{2\sin\theta\cos\theta}{r^2}\frac{\partial u}{\partial\theta}.$$

Hence finally, adding, we get

$$\frac{\partial^2 u}{\partial x^2} + \frac{\partial^2 u}{\partial y^2} = \frac{\partial^2 u}{\partial r^2} + \frac{1}{r}\frac{\partial u}{\partial r} + \frac{1}{r^2}\frac{\partial^2 u}{\partial\theta^2}.$$

The corresponding expression in three dimensions, viz.

$$\frac{\partial^2 u}{\partial x^2} + \frac{\partial^2 u}{\partial y^2} + \frac{\partial^2 u}{\partial z^2},$$

is usually denoted by $\nabla^2 u$, and differential equations containing this operator are of frequent occurrence in Higher Physics, and are known as Laplace Equations. Transformations similar to that given above frequently have to be performed. Further information on the subject of this important transformation can be found in Gibson's *Calculus*, § 99.

§ 133. Extension of Taylor's Theorem to Functions of Several Variables.

We conclude this chapter with a statement of the extension of Taylor's Theorem for more than one variable.

Consider the expression $f(x + ht,\ y + kt)$, which we will also denote by $F(t)$, and which becomes equal to $f(x + h,\ y + k)$ when $t = 1$.

Now, by Maclaurin's Theorem,

$$F(t) \equiv F(0) + tF'(0) + \frac{t^2}{2!}F''(0) + \ldots$$

Denote $\qquad ht + x$ by α,

and $\qquad\qquad kt + y$ by β.

Then $\qquad \dfrac{dF}{dt} = \dfrac{\partial F}{\partial \alpha}\dfrac{d\alpha}{dt} + \dfrac{\partial F}{\partial \beta}\dfrac{d\beta}{dt}$

$$= h\frac{\partial F}{\partial \alpha} + k\frac{\partial F}{\partial \beta}.$$

Also $\qquad \dfrac{\partial F}{\partial x} = \dfrac{\partial F}{\partial \alpha}\dfrac{\partial \alpha}{\partial x} = \dfrac{\partial F}{\partial \alpha}$

and $\qquad\qquad \dfrac{\partial F}{\partial y} = \dfrac{\partial F}{\partial \beta}.$

Hence $\qquad \dfrac{dF}{dt} = h\dfrac{\partial F}{\partial x} + k\dfrac{\partial F}{\partial y},$

or, as we may conveniently put it, the operator $\dfrac{d}{dt}$ is equivalent to the operator

$$h\frac{\partial}{\partial x} + k\frac{\partial}{\partial y}.$$

Thus $\qquad \dfrac{d^2F}{dt^2} = \left(h\dfrac{\partial}{\partial x} + k\dfrac{\partial}{\partial y}\right)\left(h\dfrac{\partial F}{\partial x} + k\dfrac{\partial F}{\partial y}\right)$

$$= h^2\frac{\partial^2 F}{\partial x^2} + 2hk\frac{\partial^2 F}{\partial x \partial y} + k^2\frac{\partial^2 F}{\partial y^2}$$

and so on.

Hence $\qquad f(x + ht, y + kt)$

$$\equiv f(x, y) + t\left\{h\frac{\partial}{\partial x} + k\frac{\partial}{\partial y}\right\}f(x, y) + \ldots$$

$$+ \frac{t^n}{n!}\left\{h\frac{\partial}{\partial x} + k\frac{\partial}{\partial y}\right\}^n f(x, y) + \ldots$$

Write $t = 1$, and we get the extended form of Taylor's Theorem, viz.

$$f(x+h, y+k) = f(x, y) + \left(h\frac{\partial f}{\partial x} + k\frac{\partial f}{\partial y}\right)$$
$$+ \frac{1}{2!}\left\{h^2\frac{\partial^2 f}{\partial x^2} + 2hk\frac{\partial^2 f}{\partial x\partial y} + k^2\frac{\partial^2 f}{\partial y^2}\right\} + \dots.$$

Owing to the resemblance between the expanded form of the operator and the Exponential Series the result is often written, (merely as an aid to the memory),

$$f(x+h, y+k) = e^{h\frac{\partial}{\partial x} + k\frac{\partial}{\partial y}} f(x, y),$$

and, in the same way, the ordinary form of Taylor's Theorem is sometimes written

$$f(x+h) = e^{h\frac{d}{dx}} f(x).$$

By putting x and y for h and k and zero for the other two elements, a series resembling that of Maclaurin may be formed. It is

$$f(x, y) = f(0, 0) + x\left(\frac{\partial f}{\partial x}\right)_0 + y\left(\frac{\partial f}{\partial y}\right)_0$$
$$+ \frac{1}{2!}\left\{x^2\left(\frac{\partial^2 f}{\partial x^2}\right)_0 + 2xy\left(\frac{\partial^2 f}{\partial x\partial y}\right)_0 + y^2\left(\frac{\partial^2 f}{\partial y^2}\right)_0\right\} + \dots.$$

These principles are all obviously capable of extension to any number of variables.

EXAMPLES XII

(1) Find $\frac{dy}{dx}$ from the relation
$$x^4 + y^4 + 4x^3y + 4xy^3 = 0.$$

(2) Find $\frac{dy}{dx}$ if $x^y + y^x = 0$.

*(3) If $(x+y)^m x^n = (x-y)^n y^m$, prove that
$$\frac{dy}{dx} = \frac{y}{x}.$$

(4) If $u \equiv x^y$, write down the total differential.

Put $x = y$ and deduce the value of $\dfrac{d}{dx} x^x$.

(5) If $f(x, y, z, w) = 0$ is an equation connecting four variables, prove that

$$(a) \quad \frac{\partial y}{\partial x} \frac{\partial x}{\partial y} = 1$$

and

$$\frac{\partial y}{\partial z} \frac{\partial z}{\partial y} = 1,$$

but

$$(b) \quad \frac{\partial x}{\partial y} \frac{\partial y}{\partial z} = -\frac{\partial x}{\partial z}$$

and

$$(c) \quad \frac{\partial x}{\partial y} \frac{\partial y}{\partial z} \frac{\partial z}{\partial w} = \frac{\partial x}{\partial w}.$$

(6) (x, y) and (r, θ) are Cartesian and polar coordinates of the same point, the origin being the pole and Ox the initial line.

Prove that there are two meanings possible for $\dfrac{\partial x}{\partial \theta}$, and that their product is r^2.

(Here again the value usually meant by $\dfrac{\partial x}{\partial \theta}$ is obtained from that equation in which x is given in terms of the polar coordinates r, θ.)

(7) If $\dfrac{\partial x}{\partial r}$, $\dfrac{\partial r}{\partial x}$, etc. bear their usual meanings, prove that

$$(a) \quad \frac{\partial x}{\partial r} = \frac{\partial r}{\partial x},$$

$$(b) \quad \frac{\partial y}{\partial \theta} = r^2 \frac{\partial \theta}{\partial y}.$$

(8) If u, a function of the Cartesian coordinates, is transformed to polar coordinates, use the total differential to prove that

$$(a) \quad \frac{\partial u}{\partial x} \equiv \cos \theta \frac{\partial u}{\partial r} - \frac{\sin \theta}{r} \frac{\partial u}{\partial \theta},$$

$$(b) \quad \frac{\partial u}{\partial y} \equiv \sin \theta \frac{\partial u}{\partial r} + \frac{\cos \theta}{r} \frac{\partial u}{\partial \theta}.$$

Hence prove that

$$\left(\frac{\partial u}{\partial x}\right)^2 + \left(\frac{\partial u}{\partial y}\right)^2 = \left(\frac{\partial u}{\partial r}\right)^2 + \frac{1}{r^2} \left(\frac{\partial u}{\partial \theta}\right)^2,$$

and deduce the conditions for a double point in polar coordinates.

(9) If in Example 8 the function is originally one of (r, θ) and is transformed to one of (x, y), find expressions for $\dfrac{\partial u}{\partial r}$ and $\dfrac{1}{r}\dfrac{\partial u}{\partial \theta}$.

(10) If $f(x, y) = 0$ where $f(x, y)$ is a homogeneous function, prove that $\dfrac{dy}{dx} = \dfrac{y}{x}$.

(11) ϕ is any function of the differences of the variables x, y, z.

Prove that $$\frac{\partial \phi}{\partial x} + \frac{\partial \phi}{\partial y} + \frac{\partial \phi}{\partial z} = 0.$$

[Hint: write $x = x_0 + t$, $y = y_0 + t$, $z = z_0 + t$, and regard as a function of t.]

(12) The coordinates (x, y) of a point become (ξ, η) after a rotation of the axes.

Given $$\left.\begin{array}{l} x = \xi \cos a - \eta \sin a \\ y = \xi \sin a + \eta \cos a \end{array}\right\}.$$

Prove that $$\left(\frac{\partial u}{\partial x}\right)^2 + \left(\frac{\partial u}{\partial y}\right)^2 = \left(\frac{\partial u}{\partial \xi}\right)^2 + \left(\frac{\partial u}{\partial \eta}\right)^2,$$

where u is any function of the coordinates.

*(13) $f(u)$ is the sum of a number of terms of the form Au^m, m being a positive or negative integer.

If $z = f\left(\dfrac{x^2}{y}\right)$, where x and y are independent variables, prove that

$$x\frac{\partial z}{\partial x} + 2y\frac{\partial z}{\partial y} = 0.$$

(14) If $x = u^2 - v^2$, $y = 2uv$, find u and v as functions of x, y, and calculate $\dfrac{\partial u}{\partial x}$.

Also find $\dfrac{\partial y}{\partial x}$ if the equations are regarded as giving u and y in terms of v and x.

(15) If u is a homogeneous function of degree n, prove that

$$x\frac{\partial^2 u}{\partial x^2} + y\frac{\partial^2 u}{\partial x \partial y} = (n-1)\frac{\partial u}{\partial x},$$

$$x\frac{\partial^2 u}{\partial x \partial y} + y\frac{\partial^2 u}{\partial y^2} = (n-1)\frac{\partial u}{\partial y},$$

and hence that

$$x^2\frac{\partial^2 u}{\partial x^2} + 2xy\frac{\partial^2 u}{\partial x \partial y} + y^2\frac{\partial^2 u}{\partial y^2} = n(n-1)u.$$

(16)
$$u = \log(x^2 + xy + y^2).$$

Prove that
$$\frac{\partial^2 u}{\partial x^2} + \frac{\partial^2 u}{\partial y^2} = \frac{\partial^2 u}{\partial x\, \partial y}.$$

(17) If h and k are constants, prove that
$$\left(h\frac{\partial}{\partial x} + k\frac{\partial}{\partial y}\right)^2 u$$

and
$$h^2\frac{\partial^2 u}{\partial x^2} + 2hk\frac{\partial^2 u}{\partial x\, \partial y} + k^2\frac{\partial^2 u}{\partial y^2}$$

are the same ; but that
$$\left(x\frac{\partial}{\partial x} + y\frac{\partial}{\partial y}\right)^2 u$$

and
$$x^2\frac{\partial^2 u}{\partial x^2} + 2xy\frac{\partial^2 u}{\partial x\, \partial y} + y^2\frac{\partial^2 u}{\partial y^2}$$

differ by
$$x\frac{\partial u}{\partial x} + y\frac{\partial u}{\partial y}.$$

(18) Transform $\dfrac{\partial^2 u}{\partial x\, \partial y}$ to polar coordinates.

*(19) If $V = (x^2 + y^2 + z^2)^{\frac{1}{2}n}$ satisfies the relation
$$\frac{\partial}{\partial x}\left(u\frac{\partial V}{\partial x}\right) + \frac{\partial}{\partial y}\left(u\frac{\partial V}{\partial y}\right) + \frac{\partial}{\partial z}\left(u\cdot\frac{\partial V}{\partial z}\right) = 0,$$

where u is a function of z only, prove that $u = Az^{-(n+1)}$, where A is an arbitrary constant.

(20) If u and v are functions of x, y connected by the relations
$$\frac{\partial u}{\partial x} = \frac{\partial v}{\partial y}, \quad \frac{\partial u}{\partial y} = -\frac{\partial v}{\partial x},$$

prove that
$$\frac{\partial^2 u}{\partial x^2} + \frac{\partial^2 u}{\partial y^2} = 0$$

and
$$\frac{\partial^2 v}{\partial x^2} + \frac{\partial^2 v}{\partial y^2} = 0.$$

(21) u and v being functions as in Example 20 and w a function expressible either in terms of x, y or of u, v, prove that
$$\frac{\partial^2 w}{\partial x^2} + \frac{\partial^2 w}{\partial y^2} = \left\{\left(\frac{\partial u}{\partial x}\right)^2 + \left(\frac{\partial v}{\partial x}\right)^2\right\}\left\{\frac{\partial^2 w}{\partial u^2} + \frac{\partial^2 w}{\partial v^2}\right\}.$$

(22) If
$$u = x^n f\left(\frac{y}{x}\right) + y^{-n} \phi\left(\frac{y}{x}\right),$$
prove that

$$x^2 \frac{\partial^2 u}{\partial x^2} + 2xy \frac{\partial^2 u}{\partial x \partial y} + y^2 \frac{\partial^2 u}{\partial y^2} + x \frac{\partial u}{\partial x} + y \frac{\partial u}{\partial y} = n^2 u.$$

(23) If $u = \phi(y + ax)$, prove that
$$\frac{\partial^2 u}{\partial x^2} = a^2 \frac{\partial^2 u}{\partial y^2},$$
and verify that the result is also true for $\psi(y - ax)$.

(24) Find $\dfrac{d^2 y}{dx^2}$ if $\dfrac{x^2}{a^2} + \dfrac{y^2}{b^2} = 1$.

(25) Find $\dfrac{d^2 y}{dx^2}$ (i) if $y^2 = 4ax$, (ii) if $xy = c^2$.

(26) Find $\dfrac{dz}{dx}$ and $\dfrac{dz}{dy}$ from the equations
$$\frac{x^2}{a^2} + \frac{y^2}{b^2} + \frac{z^2}{c^2} = 1, \quad lx + my + nz = 0.$$

(27) The side c of a triangle is calculated from measured values of a, b and C in the usual notation.

Show that the error in c due to small errors δa, δb, δC in these measurements is approximately

$$(\cos B \, \delta a + \cos A \, \delta b + ab \sin C \delta C)/c.$$

(28) Show that the error in the side c, if the formula
$$c = a \cos B + b \cos A$$
is used, may be reduced to the same expression as in Example 27.

(29) If a triangle is constructed from the data a, B, C, show that the percentage error in the value of the circum-radius may be expressed by the formula $100\left[\dfrac{\delta a}{a} + \cot A \, (\delta B + \delta C)\right]$.

(30) A triangle on measurement appears to be a right-angled triangle with the sides containing the right angle 5 and 12 feet. If measurements of length are subject to errors of 1 inch, and angular measurements to errors of 1°, prove that the possible error in the length of the hypotenuse is about 2·35 inches.

MISCELLANEOUS EXAMPLES

(1) If s_n denotes $\dfrac{1}{1^2} + \dfrac{1}{2^2} + \dfrac{1}{3^2} + \ldots + \dfrac{1}{n^2}$, show that $\operatorname*{Lt}\limits_{n \to \infty} s_n$ is a number lying between 1 and 2.

(2) Find $\operatorname*{Lt}\limits_{\theta \to 0} \theta \left(\cot \theta - \cot 2\theta \right)$.

(3) Find $\operatorname*{Lt}\limits_{x \to 0} \dfrac{1 - \cos x - x \sin x}{2 - 2 \cos x - \sin^2 x}$.

(4) Prove that $\operatorname*{Lt}\limits_{x \to 1} x^{\frac{1}{1-x}}$ is $\dfrac{1}{e}$.

*(5) Evaluate $\operatorname*{Lt}\limits_{\theta \to 0} \dfrac{1 - 2 \sin^2 \theta - \cos 3\theta}{5 \theta^2}$.

(6) Find $\operatorname*{Lt}\limits_{x \to a} \dfrac{\tan x - \tan a}{x - a}$.

(7) By consideration of $f'(x)$, show that the equation
$$f(x) \equiv x^3 + px + q = 0$$
has one real root if p is positive.

Show also that, if p is negative, there will be three real roots if
$$27q^2 + 4p^3 < 0.$$

(8) What are the directions of the curve $y = x^3 - 3x + 2$ at the points where it crosses the axes of x and y?

(9) A law for the expansion of air is $pv^{1\cdot4} = c$. If at some instant the volume V is 10 cu. feet when the pressure p is 50 lb. per square inch, find the rate at which the pressure is changing if the volume is decreasing 1 cu. ft. per second.

*(10) Find the first and second differential coefficients of
$$\log(1 + \cos x).$$

Sketch the curve $y = \log(1 + \cos x)$ between $x = 0$ and $x = 2\pi$.

(11) By considering the rate of change of $1 - \dfrac{x^2}{2} - \cos x$, show that
$$1 - \tfrac{1}{2}x^2 < \cos x < 1.$$

(12) Show, as in the previous example, that $x - \dfrac{x^3}{6} < \sin x < x$, and that
$$1 - \frac{x^2}{2} < \cos x < 1 - \frac{x^2}{2} + \frac{x^4}{24}.$$

(13) The gradient at every point on a certain curve is $\sqrt{x}$; when $x = 9$ the angle of inclination of the tangent is $45°$, and the curve passes through $(4, 2)$. Find its equation.

*(14) Find the differential coefficient of the function
$$6x^3 + 9x\,(1 + 4\cos x) - 9\sin x\,(4 + \cos x),$$
and simplify your result as much as possible.

Prove that this differential coefficient is positive.

What inferences about the original function can you draw from this result (i) when x is positive, (ii) when x is negative?

(15) Differentiate $\sqrt{1-x}\,\sin^{-1}\sqrt{x} - \sqrt{x}$.

(16) Differentiate $\sec x \tan x + \log\,(\sec x + \tan x)$.

(17) Find $\dfrac{dy}{dx}$ if $e^{x+y} + e^{x-y} = 0$.

(18) Prove that
$$\frac{d}{dx}\left\{\frac{x\sqrt{x^2+a^2}}{2} + \frac{a^2}{2}\log\,(x + \sqrt{x^2+a^2})\right\} = \sqrt{x^2+a^2}.$$

(19) Show that, if $y = Ax^3 + \dfrac{B}{x^3}$, then $x^2\dfrac{d^2y}{dx^2} = 6y$, and investigate a similar result for functions of the form $Ax^n + \dfrac{B}{x^{n-1}}$.

(20) Find $\dfrac{d^2y}{dx^2}$ if $ax^2 + 2hxy + by^2 = 1$.

(21) Show that, if $y = A\log x + B$, then
$$\frac{d^2y}{dx^2} + \frac{1}{x}\frac{dy}{dx} = 0.$$

(22) Prove that $y = Ae^{mx} + Be^{nx}$ satisfies the equation
$$\frac{d^2y}{dx^2} - (m+n)\frac{dy}{dx} + mny = 0.$$

(23) Show that, if $y = (A + Bx)\cos nx + (E + Fx)\sin nx$, then
$$\frac{d^4y}{dx^4} + 2n^2\frac{d^2y}{dx^2} + n^4 y = 0.$$

*(24) If $y = \sqrt{1+x^2}\log(x+\sqrt{1+x^2})$, prove that
$$(1+x^2)\frac{dy}{dx} = xy + 1 + x^2.$$

If y is expanded in the form $\sum\limits_{n=0}^{\infty} a_n x^n$, show that

(i) $a_{2r} = 0$; (ii) $a_1 = 1$, $a_3 = \frac{1}{3}$;

(iii) $a_{2r+1} = (-)^{r-1}\dfrac{2\,.\,4\,.\,\ldots\,(2r-2)}{3\,.\,5\,.\,\ldots\,(2r-1)}\cdot\dfrac{1}{2r+1}$ $(r>1)$.

*(25) If $y = A(\sqrt{x^2+1}+x)^n + B(\sqrt{x^2+1}-x)^n$, where A and B are any constants, prove that
$$(1+x^2)\frac{d^2y}{dx^2} + x\frac{dy}{dx} - x^2 y = 0.$$

*(26) Prove that, if $x^2 = a^2\cos^2\theta + b^2\sin^2\theta$,
$$\frac{d^2x}{d\theta^2} + x = \frac{a^2 b^2}{x^3}.$$

*(27) If $y = \sqrt{\dfrac{1-x^2}{1+x^2}}$, prove that $\dfrac{dy}{dx} = -\sqrt{\dfrac{1-y^4}{1-x^4}}$.

*(28) If $y = A + B\sin^{-1}x + (\sin^{-1}x)^2$, verify that
$$(1-x^2)\frac{d^2y}{dx^2} - x\frac{dy}{dx} = 2.$$

(29) Express $\dfrac{x^2 - 2x + 5}{(x-2)^2(x-1)}$ in the form $\dfrac{A}{(x-2)^2} + \dfrac{B}{x-2} + \dfrac{C}{x-1}$.

Hence obtain expressions for $\dfrac{dy}{dx}$ and $\dfrac{d^2y}{dx^2}$.

Write down the general expression for $\dfrac{d^n y}{dx^n}$.

(30) A chord XY is drawn parallel to a diameter AB of a circle, and a semicircle is described on XY as diameter, so that part of it lies outside the original circle.

For what position of the chord XY is the perpendicular distance between the diameter AB and the mid-point of the semicircle on XY a maximum?

(31) A circle of radius a touches the axes of coordinates at A and B, and a tangent is drawn at a point on the major arc between A and B.

Prove that the area of the triangle formed by this tangent and the axes of coordinates is a maximum if the tangent is equally inclined to the axes.

Investigate a similar problem in which the tangent is drawn at a point on the minor arc.

*(32) If a is a positive acute angle and $\sin^{-1}(\sin\theta\sin a)$ has its principal value, prove that $\theta\sin a - \sin^{-1}(\sin\theta\sin a)$ steadily increases as θ increases from 0 to $\frac{1}{2}\pi$, while a remains constant; and show that, if a and θ are both positive acute angles, the value of the expression is positive and less than

$$\tfrac{1}{2}\sqrt{(\pi^2-4)} - \cos^{-1}\frac{2}{\pi}.$$

(33) Prove that, if a and b are both positive and $a < 4b$, the function $a\sin x + b\cos 2x$ will have a maximum value for a value of x lying between 0 and $\dfrac{\pi}{2}$, and a minimum value when $x = \dfrac{\pi}{2}$.

Find these values.

(34) Prove that $\dfrac{\log_a x}{x}$ has a critical value when $x = e$, whatever the value of a, and show that it is a maximum.

*(35) A hollow cigar-shaped body is made of thin sheet metal: it consists of the curved surfaces of a circular cylinder of radius a and of two circular cones each of semi-vertical angle θ (the radii of their bases also being a), one at each end of the cylinder. The total length from vertex to vertex is b, and the thickness of the metal composing the cylindrical part is a quarter of the thickness of the metal composing the conical ends.

If a and b are given, find the value of θ which makes the total weight of the body a minimum

(i) when $b > \dfrac{2a}{\sqrt{3}}$,   (ii) when $b < \dfrac{2a}{\sqrt{3}}$.

(36) Show that the function $\dfrac{x-a}{(x-b)(x-c)}$ (a, b, c all positive) will have no maxima or minima if a lies between b and c.

(37) Show that the function $\dfrac{(x-a)(x-c)}{(x-b)(x-d)}$ will have no turning values if $a>b>c>d$. (a, b, c, d are all positive.)

(38) Discuss the variation in the value of $x\log x$ for positive values of x, and show that $x\log x=c$ has one real root if c is positive, and two real roots if c is negative and numerically less than $\dfrac{1}{e}$.

*(39) Show that $\left(\dfrac{am^2}{1+m^2},\ \dfrac{am^3}{1+m^2}\right)$ represents any point on the curve $y^2=\dfrac{x^3}{a-x}$, and find the equation of the tangent and normal at this point.

Find the coordinates of the point at which the tangent meets the curve again.

(40) Find the equation to the tangent and normal to the curve $x=a\cos^4\theta$, $y=a\sin^4\theta$, at the point of parameter θ.

Prove that the curve is part of a parabola, if θ is restricted to be real.

*(41) Find the equations of the tangent and normal at any point on the curve $x=a\cos^3 t$, $y=a\sin^3 t$, where t is a variable parameter.

Show that the axes intercept a length a on the tangent and a length $2a\cot 2t$ on the normal.

(42) In the curve of the previous example, find the values of t for a point on the curve such that the tangent at that point is also the normal at some other point on the curve.

(43) Show that two perpendicular tangents to a cycloid whose equation is $x=a(t+\sin t)$, $y=a(1-\cos t)$ intercept a constant length on the x-axis.

(44) Prove that the lengths of the subtangents to the three curves $y=4x^2$, $y=2x^3$, $y=x^4$, at the points where $x=a$, are in Harmonic Progression.

(45) Show that the curves $r = a(1 + \sin \theta)$, $r = a(1 - \sin \theta)$ meet at right angles.

(46) Find the angle between the curve $r^3 \sin 2\theta = 4$, $r^2 = 16 \sin 2\theta$.

(47) Show that the "trochoid" $x = a\theta - b \sin \theta$, $y = a - b \cos \theta$ has its maximum gradient at points where $\cos \theta = \dfrac{b}{a}$.

(48) Find where the tangent to $y^2(a - x) = x^3$ at $\left(\dfrac{a}{2}, \dfrac{a}{2}\right)$ meets the curve again.

(49) Given $y = \log \dfrac{1 + \sin x}{1 - \sin x}$, find $\dfrac{dy}{dx}$ and $\dfrac{d^2y}{dx^2}$.

(50) Find $\dfrac{d^2y}{dx^2}$ if $x^3 - xy + y^3 = 0$.

(51) Find $\dfrac{d^2y}{dx^2}$ if $x + \sin(x + y) = 0$.

(52) Find $\dfrac{dy}{dx}$ and $\dfrac{d^2y}{dx^2}$ if $x = \sin 2t$, $y = \sin^2 t$.

(53) If $y = e^{2x}(A \cos 3x + B \sin 3x)$, show that $\dfrac{d^2y}{dx^2} - 4\dfrac{dy}{dx} + 13y = 0$.

(54) Prove that $y = Ae^x + Bxe^x$ satisfies the equation
$$\frac{d^2y}{dx^2} - 2\frac{dy}{dx} + y = 0.$$

(55) If $x = \sin \theta$, transform the equation $x^2 \dfrac{d^2y}{dx^2} - x\dfrac{dy}{dx} + 2y = 0$ into a differential equation for y in terms of θ.

*(56) If $y = \cos^3 x$ and y_n denotes the nth differential coefficient of y, prove that
$$y_3 + 9y_1 + 6 \sin x = 0.$$
Deduce that, when $x = 0$, $y_1 = y_3 = y_5 = y_7 \ldots = 0$, and prove that
$$\cos^3 x = 1 - \frac{3x^2}{2!} + \frac{21x^4}{4!} - \frac{183x^6}{6!} + \ldots.$$

*(57) If $y = \tan^{-1}(1 + x)$, prove that $(x^2 + 2x + 2)\dfrac{dy}{dx} = 1$ and
$$(x^2 + 2x + 2)\frac{d^{n+2}y}{dx^{n+2}} + 2(n+1)(x+1)\frac{d^{n+1}y}{dx^{n+1}} + (n+1)n\frac{d^n y}{dx^n} = 0.$$

Expand $\tan^{-1}(1 + x)$ in a series of ascending powers of x as far as the term in x^3, and show that it contains no term in x^4.

(58) Show that the result of differentiating the equation

$$(1+x^2)\frac{dy}{dx} - 2kxy = 0, \quad n \text{ times with respect to } x,$$

is $$(1+x^2)\frac{d^{n+1}y}{dx^{n+1}} + 2(n-k)\frac{d^n y}{dx^n} + n(n-2k-1)\frac{d^{n-1}y}{dx^{n-1}} = 0.$$

*(59) Prove that

$$(x+1)\left(\frac{d}{dx}\right)^{n+1}\{(x+1)^n(x-1)^{n+1}\}$$
$$= (n+1)\left(\frac{d}{dx}\right)^n\{(x+1)^{n+1}(x-1)^n\}.$$

Prove also that $\left(\dfrac{d}{dx}\right)^n\{(x+1)^{n+1}(x-1)^n\}$ satisfies the equation

$$(1-x^2)\frac{d^2y}{dx^2} - (1+x)\frac{dy}{dx} + (n+1)^2 y = 0.$$

(60) If $\dfrac{\sin^{-1} x}{\sqrt{1-x^2}} \equiv a_0 + a_1 x + a_2 x^2 + \ldots + a_n x^n + \ldots$, prove that

(i) $a_0 = a_2 = a_4 = 0$,   (ii) $a_1 = 1$.

Find the relation between a_n and a_{n+2} and deduce that the series is $x + \dfrac{2x^3}{3} + \dfrac{2 \cdot 4}{3 \cdot 5} x^5 + \ldots$.

(61) Prove, by Maclaurin's Theorem, that

$$\tfrac{1}{2}(\sin^{-1} x)^2 = \frac{x^2}{2!} + \frac{2^2}{4!} x^4 + \frac{2^2 \cdot 4^2}{6!} x^6 + \ldots.$$

(62) Show that the expansion for $\log \cos x$ is

$$-\frac{x^2}{2} - \frac{x^4}{12} - \frac{x^6}{45} - \ldots.$$

Hence show that, as $x \to 0$, $\dfrac{\log \cos x + (1 - \cos x)}{x^4} \to -\dfrac{1}{8}$.

(63) Show, by using Maclaurin's Theorem, that, if $\dfrac{F(x)}{\phi(x)}$ assumes the form $\dfrac{0}{0}$ as x tends to zero, then its limiting value will be the same as that of $\dfrac{F'(x)}{\phi'(x)}$, unless this also assumes the form $\dfrac{0}{0}$, when the limit is that of $\dfrac{F''(x)}{\phi''(x)}$.

(64) By using the previous example, find $\underset{x \to 0}{\text{Lt}} \dfrac{e^x - 1 - x}{\cos 2x - 1}$.

(65) Find $\underset{x \to 0}{\text{Lt}} \dfrac{\log(1+x) - \sin x}{x^2}$.

(66) Given that

$$e^{x \cos a} \sin(x \sin a) \equiv x \sin a + \frac{x^2}{2\,!} \sin 2a + \ldots + \frac{x^n}{n\,!} \sin na \ldots,$$

find the sum of the series

$$\cos a + x \cos 2a + \frac{x^2}{2\,!} \cos 3a + \ldots.$$

(67) A particle moves in a straight line in such a manner that s (the distance measured from a fixed point on the line) and t (the time) are connected by the relation

$$s^2 = a^2 + b^2 t^2.$$

Find an expression for the velocity in terms of s, and prove that the acceleration varies inversely as s^3.

(68) A particle of mass m moves in a straight line so that its distance from a fixed origin at time t is $A \cos(kt + \phi)$, where A, k, ϕ are constants.

Show that the acceleration is directed towards the origin and is proportional to the distance.

Find also the maximum value of the Kinetic Energy.

(69) Two uniform rods, each of length $2l$, hinged smoothly at A, rest symmetrically over two smooth pegs at B and C (distant $2a$ apart). Show that equilibrium takes place when the rods are inclined to the horizontal at an angle a given by $\cos^3 a = \dfrac{a}{l}$.

Investigate the stability of the equilibrium, and show that the position when A is below BC is unstable.

(70) A uniform rod of length $2a$ rests in a smooth horizontal bowl of radius b of which the rim is kept vertical. Show that there is a stable position of equilibrium if $a < 2b$.

(71) The ends of a uniform rigid rod are constrained to move on two smooth wires at right angles to one another in a vertical plane Find the position of equilibrium and investigate the stability.

(72) A square $ABCD$ of equal uniform rods smoothly jointed together hangs from A, the shape being maintained by a string joining A and C. Prove by Virtual Work that the tension in the string is half the total weight of the rods.

(73) A string of length a forms the shorter diagonal of a rhombus formed of four uniform rods of length b and weight W, which are smoothly hinged together. If one of the rods is horizontal, show that the tension in the string is

$$\frac{2W.(2b^2 - a^2)}{b\sqrt{4b^2 - a^2}}.$$

(74) A particle describes the curve $x = a\cos^3\theta$, $y = a\sin^3\theta$ in such a way that θ increases at a constant rate ω radians per sec.

Show that the velocity at any point is proportional to $\sin 2\theta$, and that the tangential and normal components of acceleration are

$$3a\omega^2\cos 2\theta \quad \text{and} \quad -\frac{3a\omega^2}{4}\sin 4\theta.$$

(75) A particle describes the parabola $x = ap^2$, $y = 2ap$ in such a way that the slope of the tangent to its position at time t changes at a constant rate.

Show that the direction of the acceleration at time t is also changing at a constant rate.

(76) The coordinates of a moving point at time t are given by

$$x = Ae^{-kt}\cos pt, \quad y = Ae^{-kt}\sin pt.$$

Show that its direction of motion at time t makes with the axis of x an angle $pt - a$, where a is a constant.

Prove that the components of acceleration are

$$A(p^2 + k^2)e^{-kt}\cos(pt - 2a) \quad \text{and} \quad A(p^2 + k^2)e^{-kt}\sin(pt - 2a)$$

respectively.

(77) Find the radius of curvature at the origin for the curve

$$y^2 = \frac{ax^2}{a + x}.$$

*(78) Prove that the evolute of the two-cusped epicycloid

$$x = a(3\cos\theta - \cos 3\theta),$$
$$y = a(3\sin\theta - \sin 3\theta)$$

is the similar epicycloid

$$x = \frac{a}{2}(3\cos\theta + \cos 3\theta),$$

$$y = \frac{a}{2}(3\sin\theta + \sin 3\theta).$$

(79) Show that, for the curve

$$x = a\log(\sec t + \tan t) - a\sin t,$$
$$y = a\cos t,$$

(i) $\tan\psi = -\cot t$ and (ii) $\rho = a\tan t$.

(80) Find the coordinates of the centre of curvature at the point $\left(ct, \dfrac{c}{t}\right)$ on the hyperbola $xy = c^2$.

Hence prove that the evolute of the hyperbola is

$$(x+y)^{\frac{2}{3}} - (x-y)^{\frac{2}{3}} = (4c)^{\frac{2}{3}}.$$

(81) Find the point of contact of $y + t^2 x = at^3$ with its envelope.

Show that the radius of curvature of the envelope is

$$\pm\frac{3a}{4t}(1+t^4)^{\frac{3}{2}}.$$

(82) By considering the normal to the parabola $y^2 = 4ax$ at the point $(at^2, 2at)$, deduce that the radius of curvature of the evolute of the parabola is $6at(1+t^2)^{\frac{1}{2}}$.

(83) Show that any point on the curve $3ax^2 = y^3$ may be represented by $x = 3at^3$, $y = 3at^2$, and find the coordinates of the centre of curvature at this point.

(84) The normal at a point a of a certain curve makes intercepts $F(a)$, $f(a)$ on the axes of coordinates. Show that the coordinates of the centre of curvature at a are

$$\left\{\frac{-f'(a)(F(a))^2}{f(a)F'(a) - F(a)f'(a)}, \; \frac{F'(a)(f(a))^2}{f(a)F'(a) - F(a)f'(a)}\right\}.$$

(85) Find the envelope of the curves $y^2 = k \left\{ 1 - \left(\frac{x-\lambda}{\lambda} \right)^2 \right\}$.

Prove that it consists of the axis of y, and two straight lines parallel to the axis of x, and interpret the result geometrically.

(86) Find the envelope of the ellipse $\frac{x^2}{a^2} + \frac{y^2}{b^2} = 1$, given that $ab = c^2$.

State your result in geometrical terms.

(87) Find the envelope of the family of lines
$$ax \sec a - by \operatorname{cosec} a = a^2 - b^2.$$

(88) Show that the "tangential equation" of the ellipse $\frac{x^2}{a^2} + \frac{y^2}{b^2} = 1$ is $a^2 l^2 + b^2 m^2 = 1$.

*(89) Trace the curves
$$y^2 = 4 (x - a) (x + 1) / (x^2 - 4)$$
$$\text{(i) when } a = 1, \qquad \text{(ii) when } a = 3.$$

Show that (i) consists in part of an oval curve whose least radius vector is 1 and whose maximum radius vector is approximately 1·072, measured from the origin.

*(90) Find the asymptote of the curve
$$x^3 - y^3 = ay (x + y).$$

There are two other tangents to this curve which are parallel to the asymptote. Find their equations and the coordinates of their points of contact.

Trace the curve.

*(91) Find the asymptotes of the curve
$$(x + y) (x - 2y) (x - y)^2 + 3xy (x - y) + x^2 + y^2 = 0.$$

(92) Show that $x - y = \frac{3a}{2n}$ is an asymptote of the curve
$$x^{2n} - y^{2n} = a \{x^n y^{n-1} + 2x^{n-1} y^n\}.$$

Show also that $x + y = \pm \frac{a}{2n}$ is also an asymptote, the sign being taken as $+$ or $-$ according as n is even or odd.

(93) Find the equation of a curve which has $x=0$, $y=0$, $x+2y-4=0$ as asymptotes, and which has a double point at $(-1, 2)$.

(94) The area of a triangle is calculated by measuring its sides b, c and the contained angle A, which are 10 feet, 8 feet and 45° respectively.

If there is a possible error of 1 inch in the measurement of each side and of 1′ in the measurement of the angle, find the greatest possible error in the area.

(95) Δ is the area of a triangle expressed in terms of one side a, and two angles B and C.

Prove that $\qquad \dfrac{\partial \Delta}{\partial a} = b \sin C, \ \dfrac{\partial \Delta}{\partial B} = \dfrac{c^2}{2}.$

*(96) Given that $x = r \cos \theta$, $y = r \sin \theta$, find $\dfrac{\partial \theta}{\partial x}$, $\dfrac{\partial \theta}{\partial y}$ and $\dfrac{\partial^2 \theta}{\partial x \partial y}$, and prove that

$$\frac{\partial^{2n} \theta}{\partial x^n \partial y^n} = \frac{(2n-1)\,!}{r^{2n}} \sin \left(2n\theta - x \frac{\pi}{2} \right).$$

(97) If $z = f\left(\dfrac{x^3}{y^2}\right)$, show, by changing x into xt^2 and y into yt^3, that

$$2x \frac{\partial z}{\partial x} + 3y \frac{\partial z}{\partial y} = 0.$$

(98) If $x = \sin z$, show that the equation

$$(1-x^2)\frac{d^2y}{dx^2} - x\frac{dy}{dx} + n^2 y = 0$$

becomes

$$\frac{d^2y}{dz^2} + n^2 y = 0.$$

(99) If $f(x) \equiv ax^3 + 3bx^2 + 3cx + d$, show that the maximum and minimum values of $f(x)$ are obtained by putting $\dfrac{at - (ad - bc)}{2\,(ac - b^2)}$ instead of x in the equation $f'(x) = 0$, and solving the resulting quadratic in t.

APPENDIX

Newton's Approximation.

It has been noted on p. 96, Ex. 35, that "if $x=a$ is an approximation to a root of an equation $f(x)=0$, then $a-\dfrac{f(a)}{f'(a)}$ is possibly a closer approximation."

It does not necessarily follow that this is so, as will be seen from the following.

Suppose that $f(a)$ is small at $x=a$: that is, the point $\{a,\,f(a)\}$ on the graph $y=f(x)$ lies close to the axis of x, and, if $f(x)$ is continuous, close to a point where the curve cuts the axis.

The equation of the tangent there is $y-f(a)=f'(a)\,(x-a)$.

This meets the axis of x where $-f(a)=f'(a)\,(x-a)$,

i.e. $$x=a-\frac{f(a)}{f'(a)}.$$

The method is therefore equivalent to assuming that the point where the tangent cuts the axis is closer to the point where the curve crosses the axis than the point $\{a,\,f(a)\}$ is.

We can therefore be certain that this is a closer approximation to the root than $x=a$, if this point lies between the point where the curve cuts the axis, and the point where $x=a$.

If this is not the case, we cannot be sure that the approximation is closer, though it may be, as may be seen from a figure.

The reader will be able to see by drawing the following figures, (a) $f(a)+$, curve concave upwards, (b) $f(a)-$, curve concave downwards, (c) $f(a)+$, curve concave downwards, (d) $f(a)-$, curve concave upwards, that we can only be sure that the approximation is closer than the original value if $f(a)$ *and* $f''(a)$ *have the same sign for* $x=a$.

It should be noticed, however, that, even if we start with an unsuccessful case and repeat the process, the third approximation is closer than the second, and so on.

In practice, therefore, if the process is repeated, we get eventually a series of values which get closer and closer to the required root.

Consider the equation $\tan x = x$.

Here $f(x) \equiv \tan x - x = 0,$

$$f''(x) \equiv 2 \sec^2 x \tan x.$$

The solution clearly lies in the third quadrant and a preliminary trial by the Tables shows that it is in the neighbourhood of 258° or 4·5030 radians,

$$\tan a - a = 4 \cdot 7046 - 4 \cdot 5030 = \cdot 2016,$$

that $f(a)$ and $f''(a)$ have the same sign.

Hence $a - \dfrac{f(a)}{f'(a)}$ is therefore closer than a,

i.e. a first approximation is $4 \cdot 5030 - \dfrac{\cdot 2016}{\sec^2 a}$,

i.e. $4 \cdot 5030 - \dfrac{\cdot 2016}{(4 \cdot 8097)^2} = 4 \cdot 5030 - \cdot 0087$

$$= 4 \cdot 4943 \text{ radians.}$$

This gives $180° + 77° \ 30'$ or $257° \ 30'$ to the nearest minute.

ANSWERS TO EXAMPLES

Questions marked with an asterisk are taken from the papers of the Higher Certificate Examination and are included by permission of the Oxford and Cambridge Joint Schools Examination Board.

I (PAGE 15)

(1) $\frac{1}{2}$.   (2) $\frac{2}{3}$.   (3) $\frac{1}{4}$.   (4) $\frac{1}{2}$.   (5) $\frac{b}{a}$.

(6) $\frac{a}{b}$.   (7) $\frac{a}{b}$.   (8) $\frac{1}{2}$.   (9) 2.   (10) 1.

(11) 1.   (12) $\frac{m}{n}$.   (13) 0.   (14) 1.

II (PAGES 27–29)

(1) $2x,\ 5x^4,\ 6x^5$.   (2) $-\dfrac{1}{x^2},\ -\dfrac{2}{x^3},\ -\dfrac{3}{x^4}$.

(3) $\dfrac{1}{2\sqrt{x}},\ \dfrac{1}{3\sqrt[3]{x^2}},\ \dfrac{a}{2\sqrt{ax+b}}$.   (4) $-\dfrac{1}{2\sqrt{x^3}},\ -\dfrac{a}{2\sqrt{(ax+b)^3}}$.

(5) $8°\,8'\ (\tan^{-1}\frac{1}{7})$.   (6) $4°\,24'\ (\tan^{-1}\frac{1}{13})$ and $8°\,8'\ (\tan^{-1}\frac{1}{7})$.

(7) $71°\,36'\ (\tan^{-1}3)$.

(8) (i) $(-1, -1)$; (ii) $(-\frac{1}{2}, -\frac{3}{4})$, $(-\frac{3}{2}, -\frac{3}{4})$; (iii) $\left(\dfrac{1-2\sqrt{3}}{2\sqrt{3}}, -\dfrac{11}{12}\right)$.

(9) (i) $y+1=0$;   (ii) $x-y=\frac{1}{4}$, $x+y=-\frac{9}{4}$;

                  (iii) $y-\dfrac{x}{\sqrt{3}}+\dfrac{13-4\sqrt{3}}{12}=0$.

(10) At $(1, 0)$ and $(1, 3)$; also at $(-\frac{5}{3}, -\frac{50}{27})$ and $(-\frac{5}{3}, -\frac{85}{9})$.

(11) $3x+9y=5$.

(12) Velocity $=3\,(t-1)^2$; acceleration $=6\,(t-1)$.

(13) The motion commences at a point distant 8 units from the origin and with velocity 3. Up to 1 second there is a gradually diminishing retardation which just reduces the body to rest. It then commences to accelerate, and does so continuously.

(14) Acceleration $=2A$; initial velocity $=B$.

(15) The "gradient" of the graph, i.e. the inclination of the tangent, could be measured on the graph (having due regard to the scale). This would be the velocity.

(19) The function is decreasing for all negative values of x, and also between $x=1$, $x=2$; for all other values it is increasing.

(22) (a) 8π cu. ft. per sec.; (b) 4π sq. ft. per sec.

(23) $\tfrac{4}{3}$ sq. ft. per sec.   (24) 6 ft. per sec.

(25) $\dfrac{1}{8\pi}$ cms. per sec.   (27) At $(\tfrac{1}{2}, -\tfrac{1}{2})$ and $(\tfrac{1}{2}, \tfrac{11}{8})$; zero.

(28) $b > a^2$.   (30) $\tfrac{1}{3}$ inch per sec.

III A (PAGES 36, 37)

(2) $6x+2$.   (3) $-\dfrac{3}{x^4}+\dfrac{4}{x^3}-\dfrac{3}{x^2}$.   (4) $1-\dfrac{1}{x^2}$.

(5) $2x-1$.   (6) $2(2x-1)(x^2-x+1)$.

(7) $\dfrac{3}{2\sqrt{x}}-\dfrac{4}{3\sqrt[3]{x^2}}+\dfrac{5}{4\sqrt[4]{x^3}}$.   (8) $\tfrac{3}{2}x^{\frac{1}{2}}-1+\tfrac{1}{2}x^{-\frac{1}{2}}$.

(9) $\dfrac{1}{(x+1)^2}$.   (10) $\dfrac{9}{(4x+3)^2}$.   (11) $\dfrac{x^3+8x+13}{(x+4)^2}$.

(12) $\cos x - x\sin x$.   (13) $-\dfrac{\cos x + x\sin x}{x^2}$.

(14) $2x\sin x + x^2\cos x$.   (15) $\sin 2x$; $-\sin 2x$.

(16) $\dfrac{x\cos x - 3\sin x}{x^4}$.   (17) $2\tan x\sec^2 x$; $2\tan x\sec^2 x$.

(18) $x(\cos^2 x - \sin^2 x)+\sin x\cos x$.   (19) $3\sin^2 x\cos x$.

(20) $n\sin^{n-1} x\cos x$.   (21) $4x(x^2+1)$.

(22) $\dfrac{2x\cos x+(x^2+1)\sin x}{\cos^2 x}$.   (23) $e^x\left\{\dfrac{\sin x-\cos x}{\sin^2 x}\right\}$.

(24) $e^x\dfrac{1+\cos x}{1-\sin x}$.   (25) $\tfrac{1}{2}\cos x$.   (26) $-\dfrac{3\sin x}{(2+\cos x)^2}$.

(27) $\dfrac{6(1-\cos x)^3}{(9+6\cos x)^2}$.

III B (Pages 42, 43)

(1) $6(1+2x)^2$. (2) $12x(1+2x^2)^2$. (3) $2nbx^{n-1}(a+bx^n)$.

(4) $mnbx^{n-1}(a+bx^n)^{m-1}$. (5) $\dfrac{6}{(1-3x)^3}$. (6) $3\cos 3x$.

(7) $3\cos(3x-2)$. (8) $3\cos(x-2)\sin^2(x-2)$.

(9) $2xe^{x^2}$. (10) $2e^{2x}$. (11) $4xe^{2x^2}$.

(12) $-\dfrac{x}{\sqrt{a^2-x^2}}$. (13) $\dfrac{3+10x}{2\sqrt{1+3x+5x^2}}$.

(14) $\dfrac{b+2cx}{2\sqrt{a+bx+cx^2}}$. (15) $\dfrac{b+2cx}{2\sqrt{a+bx+cx^2}}e^{\sqrt{a+bx+cx^2}}$.

(16) $\dfrac{2x^2+ax+a^2}{\sqrt{x^2+a^2}}$. (17) $\dfrac{1-x}{(x^2+1)^{\frac{3}{2}}}$. (18) $\dfrac{x^2-3x^{-4}}{(x^3+3x^{-3})^{\frac{2}{3}}}$.

(19) $\dfrac{e^x}{\sqrt{1-e^{2x}}}$. (20) $\cot x$. (21) $\cot x+\tan x$.

(22) $-\dfrac{1}{(1+x)\sqrt{1-x^2}}$. (23) $-\dfrac{1}{x^2}+\dfrac{1}{x^3\sqrt{1-x^2}}$.

(24) $\dfrac{1}{\sqrt{x^2+1}}$. (25) $\dfrac{2x+1}{x^2+x+1}\log_{10}e$.

(26) $\dfrac{1}{\sqrt{1-x^2}}e^{\sin^{-1}x}$. (27) $\left(\tan^{-1}x+\dfrac{x}{1+x^2}\right)e^{x\tan^{-1}x}$.

(28) $\dfrac{x}{(x^2+1)\sqrt{\log(x^2+1)}}$. (29) $\dfrac{1}{x\log x}$. (30) $\dfrac{\sqrt{2}(1+x^2)}{1+x^4}$.

(31) $\dfrac{1}{\cos\frac{x}{2}(1+e)+\sin\frac{x}{2}(1-e)}\sqrt{\dfrac{1-e^2}{8\sin x}}$. (32) $\dfrac{\sqrt{2}(1-x^2)}{1+x^4}$.

(33) $\sec x$. (34) $\dfrac{a-b}{2(2x-a-b)\sqrt{(x-a)(x-b)}}$.

(35) $\dfrac{1}{\sqrt{a^2-x^2}}$. (36) $\dfrac{a}{x^2+a^2}$. (37) $\dfrac{1}{x\sqrt{x^2-1}}$.

(38) $\dfrac{2}{\sqrt{1-x^2}}$. (39) $-\dfrac{2(x^2+2)}{x^4+4}$. (40) $\dfrac{1}{\sqrt{x^2-1}}$. (41) $\dfrac{1}{\sqrt{1+x^3}}$.

(42) $\dfrac{\sqrt{1-x^2}}{1+x^3}$. (43) $3ax^3+bx$. (44) $\dfrac{1-e^{-2x}}{2}$.

III C (PAGES 47, 48)

(1) $2x \cdot 2^{x^2} \cdot \log_e 2.$

(2) $x^x \{\log_e x + 1\}.$

(3) $\dfrac{x e^x \cdot e^x}{x} \{x \log x + 1\}.$

(4) $x^{ex} \cdot e \{\log x + 1\}.$

(5) $x^x \{\log x + 1\} + x^{\frac{2}{x}} \cdot \dfrac{2}{x^2} \{1 - \log x\}.$

(6) $\dfrac{2x}{a^2} - \dfrac{2x^3}{a^2 \sqrt{x^4 - a^4}}.$

(7) $\dfrac{2}{x-1} - \dfrac{1}{x+3} - \dfrac{1}{x+1}.$

(8) $-\dfrac{1}{\sqrt{1-x^2}}.$

(9) $-\dfrac{2}{\sqrt{1-x^2}}.$

(10) $\dfrac{3}{\sqrt{1-x^2}}.$

(11) $\dfrac{3}{1+x^2}.$

(12) $4x + \dfrac{4x^3 - 2}{\sqrt{x^2 - 1}}.$

(13) $-1.$

(14) $-\dfrac{x+3}{y+4}.$

(15) $-\dfrac{b^2 x}{a^2 y}.$

(16) $-\dfrac{ax+hy+g}{hx+by+f}.$

(17) $-\dfrac{y^{\frac{1}{3}}}{x^{\frac{1}{3}}}.$

(18) $-\dfrac{1}{x(\log y + 1)}.$

(19) $-\dfrac{\sin y + y \cos x}{x \cos y + \sin x}.$

(20) $\dfrac{y^2}{x - xy \log x}.$

(21) $-\dfrac{y}{x}.$

(22) $\dfrac{y}{x}.$

(23) $2(ax^2 + 2hxy + by^2).$

(24) $\tfrac{3}{2}(x+y)(x^{\frac{1}{2}} + y^{\frac{1}{2}}).$

(25) $4(x^4 + x^3 y + 4x^2 y^2 + xy^3 + y^4).$

(26) $0.$

(28) $\dfrac{3(\cos^2 \theta - \sin^2 \theta)}{\sin \theta}.$

(29) $\dfrac{3}{4at}.$

IV (PAGES 61–63)

(1) Parts are equal.

(2) $\pm \sqrt{a^2 + b^2}.$

(3) Maximum 2; minimum $\tfrac{6}{5}$.

(4) Minima occur when $x = 2n\pi + \dfrac{5\pi}{6}$; maxima when $x = 2n\pi + \dfrac{\pi}{6}$.

(5) A square.

(6) The side of the square must be $7 - \sqrt{13}$ inches.

(7) If a is radius of sphere, the height of maximum cylinder $= \dfrac{2a}{\sqrt{3}}$; that of maximum cone $= \dfrac{4a}{3}$.

(8) $3(20)^{\frac{2}{3}}\pi^{\frac{1}{3}}$ square feet.

(9) Width of rectangular opening $=\dfrac{4(4+\sqrt{3})}{13}$ feet.

(10) Point is $\left(\sqrt{\dfrac{33}{14}},\ \sqrt{\dfrac{77}{6}}\right)$.   (11) $x=2$.

(15) $a+b+2\sqrt{ab}$.

(16) $\dfrac{\pi}{4}$ and $\tan^{-1}2$. (The first gives a maximum, the second a minimum.)

(17) (i) $\dfrac{r}{h}=\dfrac{1}{\sqrt{2}}$; (ii) $\dfrac{r}{h}=\dfrac{1}{2\sqrt{2}}$.

(19) Minimum value, $y=-24$, occurs where $x=-2$. The value $x=1$ gives an inflexion on the curve.

(20) Critical value $x=\dfrac{a-b-1}{a-b+1}$: Max. if $a>b+4$; Min. if $a<b+4$; Inflexion if $a=b+4$; discontinuity if $a=b$.

(21) Max. 1; Min. 9.

(22) The maximum volume $=\dfrac{s^3}{1350}$ ($s=$ length of string).

(23) (i) $\dfrac{ab}{4}$; (ii) $\dfrac{a^2b^2}{a^2+b^2}$.

(25) Least focal chord is latus rectum.   (26) $\dfrac{d}{\sqrt{a^2+b^2+c^2}}$.

V A (Pages 70–72)

(1) $y-8x+1=0$; $8y+x-122=0$.

(2) $y-9x-27=0$; $9y+x+3=0$.

(3) $\pm 4y+3x-25=0$; $\mp 3y+4x=0$.

(4) $\pm y=x+a$; $\pm y+x=3a$.

(5) $\dfrac{x}{a}+\dfrac{y}{b}=\sqrt{2}$; $ax\sqrt{2}-by\sqrt{2}=a^2-b^2$.

(6) $2y+7x-3=0$; $14y-4x-21=0$.

(7) $y\cos\dfrac{t}{2}+x\sin\dfrac{t}{2}=at\sin\dfrac{t}{2}+2a\cos\dfrac{t}{2}$;

$y\sin\dfrac{t}{2}-x\cos\dfrac{t}{2}+at\cos\dfrac{t}{2}=0$.

(8) $x(x_0^3-y_0)+y(y_0^3-x_0)-x_0y_0=0$;

$x(y_0^3-x_0)-y(x_0^2-y_0)+(x_0-y_0)(x_0y_0+x_0+y_0)=0$.

(9) $x(ax_0+hy_0)+y(hx_0+by_0)-1=0$;

$x(hx_0+by_0)-y(ax_0+hy_0)-h(x_0^2-y_0^2)+(a-b)x_0y_0=0$.

(10) $xy_0 + x_0y - 2c^2 = 0$; $xx_0 - yy_0 + y_0^2 - x_0^2 = 0$.

(11) $ty = x + at^2$; $y + tx = 2at + at^3$.

(12) At $x = a$

$$y - \frac{c}{2}\left(e^{\frac{a}{c}} + e^{-\frac{a}{c}}\right) = \frac{1}{2}\left(e^{\frac{a}{c}} - e^{-\frac{a}{c}}\right)(x-a)\,;$$

$$\left[y - \frac{c}{2}\left(e^{\frac{a}{c}} + e^{-\frac{a}{c}}\right)\right]\left(e^{\frac{a}{c}} - e^{-\frac{a}{c}}\right) + 2(x-a) = 0.$$

(13) $2t_0 y + (1 - 3t_0^2)\,x - (t_0^2 - 1)^2 = 0$.

(14) Tangent at $(at^2,\ at^3)$ is $2y - 3tx + at^3 = 0$.

(15) $6ayy_0 = (3x_0^2 + 2ax_0)\,x - x_0^3$ is the tangent at $(x_0,\ y_0)$.

(16) $2xx_0(x_0 - 1) - 2yy_0 - (x_0^2 + y_0^2) = 0$.

(17) $y(1 - 2m^3) - x(2m - m^4) + am^2 = 0$.
The asymptote $3x + 3y + a = 0$ is deduced by putting $m = -1$.

(18) Subtangent $= \dfrac{a\sin^2\theta}{\cos\theta}(2 - \sin^2\theta)$.

Subnormal $= a\cos\theta\,(2 - \sin^2\theta)$. $p = a\sin^2\theta$.

(19) $\cos\dfrac{x}{c}$; $\sin\dfrac{x}{c}$.   **(21)** $\dfrac{1}{b'} - \dfrac{1}{b} = \dfrac{1}{a} - \dfrac{1}{a'}$.

(23) The question can be done best by putting $x = a\sin^2\theta$. The subtangent is then $2a\cos^2\theta\sin^2\theta$ and is a maximum if $\theta = \dfrac{\pi}{4}$.

(25) The parabolas cut at an angle $\tan^{-1}\frac{3}{4}$ and the tangent to the folium bisects the exterior angle between the tangents to the parabola.

V B (Pages 78, 79)

(1) $Ahx^2 + Bky^2 = 1$, if (h, k) is the fixed point.

(2) Points of contact where $x = 2,\ 6,\ -\frac{3}{2}$.

(3) $3hx^2 - ay^2 - 2aky = 0$.   **(6)** $\phi = a$; $p = ae^{\theta\cot a}\sin a$.

(8) $2ap^2 = r^3$.   **(12)** $p = \dfrac{a^2}{\sqrt{a^2 + b^2}}$.   **(13)** $ake^{\kappa\theta}$.

(14) $r\cos(\theta + a) = a\sqrt{\cos 2a}$.

(15) $p = a(\cos 2a)^{\frac{3}{2}}$ at $\theta = a$; tangent is $r\cos(\theta - 3a) = a(\cos 2a)^{\frac{3}{2}}$.

VI (PAGES 94–96)

(1) $\dfrac{(n+r)\,!}{r\,!}\,x^r.$ (2) $(-1)^n\,\dfrac{(n+r-1)\,!}{(r-1)\,!}\,x^{-(n+r)}.$

(3) $(-1)^{n-1}\dfrac{(n-1)\,!}{x^n}.$ (4) $(-1)^n\dfrac{(n+1)\,!}{(x+1)^{n+2}}.$

(5) $(-1)^n\dfrac{n\,!}{2}\left\{\dfrac{1}{(x+1)^{n+1}}-\dfrac{1}{(x+3)^{n+1}}\right\}.$

(6) $\cos\left(x+\dfrac{n\pi}{2}\right).$ (7) $2^{n-1}\sin\left(2x+\dfrac{n-1}{2}\,\pi\right).$

(8) $2^{\frac{n}{2}}e^x\cos\left(x+\dfrac{n\pi}{4}\right).$

(9) $\tfrac{1}{2}e^x\{1+5^{\frac{n}{2}}\cos\left(2x+n\tan^{-1}2\right)\}.$

(10) $\dfrac{1}{4}\left\{3\sin\left(x+\dfrac{n\pi}{2}\right)-3^n\sin\left(3x+\dfrac{n\pi}{2}\right)\right\}.$

(11) $\dfrac{(-1)^n\,n\,!}{2a}\left\{\dfrac{1}{(x-a)^{n+1}}-\dfrac{1}{(x+a)^{n+1}}\right\}.$

(12) $(-1)^n n\,!\left\{\dfrac{1}{(x+1)^{n+1}}+\dfrac{1}{(x-1)^{n+1}}-\dfrac{1}{x^{n+1}}\right\}.$

(13) $e^x\{x^3+2nx+n\,(n-1)\}.$

(14) $\cos\left(x+\dfrac{n\pi}{2}\right)\{x^3-3n\,(n-1)\,x\}$
$$+\sin\left(x+\dfrac{n\pi}{2}\right)\{3nx^2-n\,(n-1)\,(n-2)\}.$$

(15) $\dfrac{2\,(-1)^{n-1}\,(n-3)\,!}{(1+x)^{n-2}}.$

(16) $x^3\dfrac{d^{n+2}y}{dx^{n+2}}+2\,(n-1)\,x\,\dfrac{d^{n+1}y}{dx^{n+1}}+(2n^2-3n+2)\dfrac{d^n y}{dx^n}.$

(17) $e^x\{(1-{}^nC_2+{}^nC_4-\ldots)\sin x+({}^nC_1-{}^nC_3+\ldots)\cos x\}.$

(20) $1+x+\tfrac{1}{2}x^2-\tfrac{1}{6}x^3-\tfrac{7}{24}x^4+\tfrac{1}{24}x^5+\ldots.$

(21) $1-\dfrac{x^2}{2\,!}+\dfrac{x^4}{4\,!}-\dfrac{x^6}{6\,!}+\ldots.$

(22) $x-\dfrac{x^3}{3}+\dfrac{x^5}{5}-\ldots.$ {x numerically <1.}

(23) $2\left\{x+\dfrac{x^3}{3}+\dfrac{x^5}{5}+\ldots\right\}$.   {$x$ numerically <1.}

(24) $x+\dfrac{x^3}{3\,|}+\dfrac{x^5}{5\,|}+\ldots$.

(25) $x-\dfrac{1}{2}\dfrac{x^3}{3}+\dfrac{1.3}{2.4}\dfrac{x^5}{5}-\ldots$.   {$x$ numerically <1.}

(29) (i) $\dfrac{\pi}{4}x-x^2+\dfrac{\pi}{12}x^3$.   (ii) $y-2y^2+2y^3$. $\left\{y=\dfrac{\pi}{4}-x.\right\}$

(33) [Note. The function given clearly vanishes when $x=a$ and when $x=a+h$. Hence its derivative vanishes at some intermediate point, provided $f(x)$, $f'(x)$ are not discontinuous. Hence the theorem follows. Similarly Higher Theorems of the Mean can be deduced; and in general the theorem of the nth order will imply the existence of the first n derivatives, and the continuity of all of these but the last.]

(35) $3\cdot032$.

VII (Pages 108–110)

(7) The equilibrium is stable as long as the height h of the cone <4 times the radius of the hemisphere. The limiting case when $h=4R$ is unstable.

(8) The limiting case is unstable.   (10) $\frac{2}{3}W$ in each rod.

(11) $3\sqrt{3}\,W$.   (12) $4W$.

(16) Hint: The direction of the accel. gives us $y\dfrac{d^2x}{dt^2}-x\dfrac{d^2y}{dt^2}=0$:

this is seen to result from differentiating $y\dfrac{dx}{dt}-x\dfrac{dy}{dt}=$ constant.

(17) $W\dfrac{2\sin 72°+\sin 36°}{\cos 72°+\cos 36°}$, or $W\dfrac{1+\sin 18°}{\cos 54°}$.

(18) $3W$ (W is the weight of a rod).

VIII A (Pages 120, 121)

(1) $4a\cos\psi$.   (2) $ae^{-\frac{s}{a}}\cos\psi$.   (3) $c\cosh^2\dfrac{x}{c}$.

(4) $c\sec\dfrac{x}{c}$.   (5) $4a\cos\dfrac{\theta}{2}$.   (6) $2a(1+t^2)^{\frac{3}{2}}$.

(7) $\dfrac{(a^2\sin^2\theta+b^2\cos^2\theta)^{\frac{3}{2}}}{ab}$.   (8) $\sqrt{A^2+B^2}$.   (9) $\dfrac{4a}{3}\cos\dfrac{\theta}{2}$.

(10) $\dfrac{3a}{2}$. (N.B. There are two branches to this curve at O, as we see in Ch. IX, but they are equal; hence their radii of curvature are equal.)

(11) $\dfrac{(b^2 - a^2)^{\frac{1}{2}}}{2}$.   **(12)** ρ at the point $(r, \theta) = \dfrac{a^n}{(n+1)\, r^{n-1}}$.

(13) $\dfrac{(1 - 2t^2 + 9t^4)^{\frac{3}{2}}}{2\,(1 + 3t^2)}$.

(19) In the case $\dfrac{a}{A} = \dfrac{b}{B} = \dfrac{c}{C}$ the curve degenerates into a straight line, and every point obeys the condition for inflexion.

(20) The inflexion occurs where $x = -\dfrac{b}{3a}$.

VIII B   (Pages 127, 128)

(1) $\rho = \dfrac{(a^2 \sin^2 \theta + b^2 \cos^2 \theta)^{\frac{3}{2}}}{ab}$; centre is

$$\left[\frac{a^2 - b^2}{a} \cos^3 \theta, \; -\frac{a^2 - b^2}{b} \sin^3 \theta \right].$$

(2) If the coordinates $x = a \cos \theta$, $y = b \sin \theta$ are substituted in the equation of any circle, an equation in θ results, giving the eccentric angles of the four intersections; the sum of these angles can be shown to be $2n\pi$. If the circle is the circle of curvature at θ, three of these intersections coincide at θ, and the rest of the example follows.

(3) The corresponding result for the parabola is that if a circle and a parabola meet in 4 points, the sum of their ordinates is zero. The centre of curvature is $(2a + 3at^2, \; -2at^3)$, with the usual notation.

(7) $8ay + a^2 = 3x^2 + 6ax$.   **(8)** $3\,(x^2 + y^2) + 2\,(y - x) = 0$.

(9) $2\,(2y - 5)^3 = 27\,(x - 5)^2$.

(10) The parametric equation to the evolute is

$$x = \frac{t^4 + 3t + 3}{t}, \; y = \frac{3t^4 - 5t^3 + 1}{t^3}.$$

(13) $r^2 + 2e^{\frac{\pi}{2}}\, r \cos \theta - e^{\pi} = 0$. $\{$Take $r^2 + r\,(A \cos \theta + B \sin \theta) + c = 0$ and equate r, $\dfrac{dr}{d\theta}$, and $\dfrac{d^2 r}{d\theta^2}$ for the two curves.$\}$

(14) Find the "straight line of closest contact," it is

$$r \sin \theta = \frac{a}{2\sqrt{2}}.$$

(15) $\dfrac{2\pi - a}{3}$, $\dfrac{4\pi - a}{3}$, $\dfrac{6\pi - a}{3}$.

(18) $ty + t^3 x = c(1 + t^4)$: use result of No. 16.

IX (PAGES 138, 139)

(1) $x^2 + 4ay = 0$. (2) $x^2 + y^2 = a^2$.

(3) $x^2 y^2 = a^2 (x^2 + y^2)$. (4) $x^2 + y^2 = ax$.

(5) $\sqrt{cx} + \sqrt{cy} = 1$ or $c^2 (x - y)^2 - 2c(x + y) + 1 = 0$.

(6) $x^{\frac{2}{3}} + y^{\frac{2}{3}} = c^{\frac{2}{3}}$. (7) The four lines $\pm x \pm y = c$.

(8) The parabola $y = \dfrac{u^2}{2g} - \dfrac{gx^2}{2u^2}$.

(9) (i) The envelope of a line which cuts off from the axes intercepts whose sum is constant is a parabola touching the axes of coordinates (more accurately: 4 parabolas, one in each quadrant).

(ii) The envelope of a family of ellipses, the sum of the squares of whose semi-axes is constant, consists of a square whose sides are inclined to the axes at 45°.

(10) $a^2 (l^2 + m^2) = 1$. (11) $am^2 = l$.

(12) $4al^3 + 27m^2 = 0$. (13) $4c^2 lm = 1$.

(14) $\sqrt{al} + \sqrt{am} + 1 = 0$. (15) $\dfrac{1}{al} + \dfrac{1}{bm} + 1 = 0$.

(16) $\left(\dfrac{x}{a}\right)^{\frac{2}{3}} + \left(\dfrac{y}{b}\right)^{\frac{2}{3}} = 1$.

(17) $(ax)^{\frac{2}{3}} + (by)^{\frac{2}{3}} = (a^2 - b^2)^{\frac{2}{3}}$; centre

$$\left(\frac{a^2 - b^2}{a} \cos^3 \theta, \quad -\frac{a^2 - b^2}{b} \sin^3 \theta\right).$$

(18) $9c^2 \{16c^2 x - x^2 y - 3y^3\} \{16c^2 y - xy^2 - 3x^3\}$
$$= 4 (4c^2 - xy)^2 (8c^2 + xy)^2 ;$$

centre $\left\{\dfrac{c}{2t^3}(3t^4 + 1), \ \dfrac{c}{2t}(t^4 + 3)\right\}.$

(19) $2(2x-1)^3 + 27(y-4)^2 = 0$; centre $\left\{\dfrac{1-6t^2}{2},\ 4(1+t^3)\right\}$.

(20) The evolute is given parametrically by the equations

$$x = -(1 - \cos t); \quad y = t + \sin t.$$

(23) (i) The tangent at the vertex.   (ii) $r\cos\theta + a\sin^2\theta = 0$.

(24) The auxiliary circle.

(27) $x\sin 2\theta - y\cos 2\theta = a\sin\theta$.

The caustic is $(x^2 + y^2) = \dfrac{a^2}{4}\left\{3\left(\dfrac{y}{a}\right)^{\frac{2}{3}} + 1\right\}$.

(28) The reflected ray is $x\cos\theta + y\sin\theta = a\{\sin\theta + \theta\cos\theta\}$. The elimination of θ between this and its derivative leads to

$$a\cos^{-1}\left(\dfrac{y-a}{a}\right)^{\frac{1}{2}} = x \pm [(y-a)(2a-y)]^{\frac{1}{2}}.$$

X A (Pages 148, 149)

In the following sets of examples the equation to the branch only is given; the reader should, of course, sketch the form for himself.

(5) $y = \dfrac{x}{2} - \dfrac{x^3}{48a^2}$.   (6) $y = x + \dfrac{2x^3}{a}$.

(7) (i) $y = 3x + \dfrac{75x^2}{4}$.   (ii) $y = \dfrac{x}{3} - \dfrac{65x^2}{108}$.

(8) (i) $y = x + \dfrac{x^2}{2}$.   (ii) $y + x = \dfrac{x^2}{6}$.   (iii) $y = \dfrac{x}{2} - \dfrac{2x^2}{3}$.

(9) $y^2 = \pm\sqrt{5}ax$.   (10) $4ay = 3x^2$.

(11) $y - x = \pm 2\sqrt{\dfrac{x^3}{a}}$.

(12) (i) $x = \dfrac{y^3}{2}$.   (ii) $y = \dfrac{x^2}{2}$.   (iii) $y = x + \dfrac{x^2}{2}$ and $y = x - x^2$.

(13) (i) $y^3 = 5x^2$.   (ii) $x^3 = 5y^2$.

(14) First approx. $y = x^2$; second $y - x^2 = \pm\sqrt{2x^5}$, i.e. the branch is a rhamphoid cusp in the first quadrant.

(15) $a^2 y = x^3$.

(16) First approx. $y^2 = -x$; but if $y^2 + x$ is put equal to β we find $\beta = \sqrt{-y^5}$; which gives a rhamphoid cusp touching the lower limb of the first approximation. The other tangent gives $y + x = x^3$.

(17) The rationalized form of the equation, viz.
$$x^2 - 2xy + y^2 - 2a(x+y) + a^2 = 0,$$
should be used.

(18) The rationalized form is $(x^2 + y^2 - a^2)^3 = 27a^2x^2y^2$.

(19) Cartesian equation is $\{x^2 + y^2 - ax\}^2 = a^2x^3 + a^2y^2$.

(20) (i) $u_1 + \underset{u_1 \to 0}{\text{Lt}} \{u_3\} = 0$.

   (ii) Two branches (a) $u_1 + \underset{u_1 \to 0}{\text{Lt}} \{u_2\} = 0$.

             (b) $u_1 + \underset{u_1 \to 0}{\text{Lt}} \left\{\dfrac{u_4}{u_2}\right\} = 0$.

   (iii) Two branches (a) $u_1 + \underset{u_1 \to 0}{\text{Lt}} \left\{\dfrac{u_4}{v_1}\right\} = 0$.

             (b) $v_1 + \underset{v_1 \to 0}{\text{Lt}} \{v_2\} = 0$.

(21) Inflexion at each point.

(22) The question assumes $a < b$. The origin is a node and the shape at the other points parabolic.

X B (Page 153)

(1) Acnode at $(0, 0)$.

(2) A cusp at 0; the tangent being $x = y$.

(3) $(0, 0)$ is a node.   (4) Cusps at $(\pm a, 0)$ and $(0, \pm a)$.

(5) Node at $(0, 0)$.   (6) Inflexion given by $t^3 = 4$.

(7) Inflexions where $t = 0$, $t = -1$. Acnode at $(0, 1)$, given by
$$t^2 + t + 1 = 0.$$

(8) Cusp at 0; tangent $x = 0$.

XI (Pages 164, 165)

(1) $3x + 4y = \frac{2 \cdot 2}{5}$.

(2) $2y + 3 = 0$; $y - x = 1$; $y - x = 2$; $y - 2x = -\frac{3}{2}$.

(3) $y = \pm\sqrt{3}$; $y - x = \pm\sqrt{3}$.   (4) $y - x = \frac{3}{2}$; $y + x = 0$.

(5) $x = 0$; $y + x = 0$; $y - x = 0$.

(6) $y - x = -1$; $y + x = 0$.

(7) $x + 1 = 0$; $y = x$; $y + x = 0$.

(8) $x = 4$; $y - x = 3$; $y + x = 3$.

(9) $x = 0$; $y + 2 = 0$; $x + 2y = -2$.

(10) $y - x = 0$.

(11) $x + \dfrac{3a}{2} = 0$; $x - y = a$; $x - y = 2a$.   (12) $y = x$; $y + x = 0$.

(13) Parabolic branch parallel to $x = 0$; $x + y = a$.

(14) $x = 0, y = 0$.   (15) $x = 0, y = 0, x = y, x + y = 0$.

(16) $x = \pm \sqrt{2}a$; $x - y = -a$; $x + y = -a$.

In the following examples a few hints to the shape of the curve are given, it being impossible to reproduce each diagram.

(17) One asymptote parallel to $x + y = 0$; the origin a node, with the axes as nodal tangents. Line $y = x$ cut at O and at $\left(\dfrac{3a}{2}, \dfrac{3a}{2} \right)$. The nodal branches form a loop.

(18) The origin a double cusp ; the cuspidal tangents being the axes, the branches forming a loop. One asymptote parallel to $x + y = 0$. The asymptote is cut in two real points in the first quadrant.

(19) The asymptote is given in No. 1 and the shape at O in No. 7, X A. Asymptote is not cut except at infinity.

(20) Cusps at (a, a), $(a, -a)$. The curve is closed. Shape at O parabolic.

(21) $y + 2x = 0$ is an asymptote and is only cut at O. $y + x = 0$ is an inflexional tangent at O.

(22) There are two asymptotes parallel to $x = y$, and $y + 2x = 0$ is also an asymptote. $y + x = 0$ is an inflexional tangent at O. One branch lies between the parallel asymptotes and goes through O. There is another branch between each of the parallel asymptotes and the third. Each axis is cut twice ; the asymptotes not at all.

(23) Similar to No. 22 ; but instead of $y + 2x = 0$, $y = 0$ is an asymptote and is not cut.

(24) The origin is a node, the branches forming a loop in the third quadrant. One asymptote parallel to $x + y = 0$; it is cut once.

(25) The axis of x is an asymptote with two branches approaching its negative extremity. Two asymptotes, parallel to $y=x$, $y+x=0$, the latter being cut twice. The origin is an inflexion, and there is a branch cutting Oy below O.

(26) Two asymptotes parallel to each axis. The origin a node; neither axis cut except at O, and the curve is symmetrical with respect to Ox. No branch between $x=1$ and $x=2$.

(27) O is an inflexion, with the axis of y as tangent. There are two asymptotes parallel to $y=x$. One branch lies between the asymptotes, passing through the origin. There are two other branches lying between the oblique asymptotes and the axis of x, which is also an asymptote.

(28) Each axis is an inflexion at O, and $y=x$, $y+x=0$ are asymptotes. The curve consists of four equal parts, each lying between an asymptote and the axis. No line crossed except at O.

(29) Two asymptotes parallel to each axis. Each one is cut in one point. Ox is an inflexion on one branch which lies between the asymptotes parallel to Ox. A branch approximating to the curve $y=x^3$ lies between the asymptotes parallel to Oy, and the curve is completed by two branches in the second and fourth quadrants.

XII (Pages 181–185)

(1) $-\dfrac{x^3+3x^2y+y^3}{x^3+3xy^2+y^3}$ or $\dfrac{y}{x}$. (2) $-\dfrac{y\,(x^{y-1}+y^{x-1}\log y)}{x\,(x^{y-1}\log x+y^{x-1})}$.

(3) Use Euler's Theorem for the function $(x+y)^m x^n-(x-y)^n y^m$.

(4) $\dfrac{du}{dt}=yx^{y-1}\dfrac{dx}{dt}+y^x\log y\,\dfrac{dy}{dt}$; $\dfrac{d}{dx}x^x=x^x+x^x\log x$.

(9) $\dfrac{\partial u}{\partial r}=\cos\theta\,\dfrac{\partial u}{\partial x}+\sin\theta\,\dfrac{\partial u}{\partial y}$; and $\dfrac{1}{r}\dfrac{\partial u}{\partial\theta}=-\sin\theta\,\dfrac{\partial u}{\partial x}+\cos\theta\,\dfrac{\partial u}{\partial y}$.

(14) u and v are given by

$$u^2=\frac{x+\sqrt{x^2+y^2}}{2},\quad v^2=\frac{-x+\sqrt{x^2+y^2}}{2},$$

and
$$\frac{\partial u}{\partial x}=\frac{u}{2\sqrt{x^2+y^2}}.$$

Also $y^2=2v\sqrt{v^2+x}$, whence $\dfrac{\partial y}{\partial x}=\dfrac{2v^2}{y^3}$.

(18) $\cos\theta\sin\theta\left\{\dfrac{\partial^2 u}{\partial r^2}-\dfrac{1}{r^2}\dfrac{\partial^2 u}{\partial\theta^2}\right\}+\dfrac{\cos^2\theta-\sin^2\theta}{r}\dfrac{\partial^2 u}{\partial r\,\partial\theta}$

$$+\dfrac{\sin^2\theta-\cos^2\theta}{r^2}\dfrac{\partial u}{\partial\theta}-\dfrac{\sin\theta\cos\theta}{r}\dfrac{\partial u}{\partial r}.$$

(24) $\dfrac{b^4}{a^2 y^3}.$ (25) (i) $-\dfrac{4a^2}{y^3}$; (ii) $\dfrac{2c^3}{x^3}.$

(26) $\dfrac{dz}{dx}=\dfrac{c^2\{b^2 mx-a^2 ly\}}{a^2\{c^2 ny-b^2 mz\}}$ and $\dfrac{dz}{dy}=\dfrac{c^2\{a^2 ly-b^2 mx\}}{b^2\{c^2 nx-a^2 lz\}}.$

MISCELLANEOUS EXAMPLES (PAGES 186–197)

(2) $\frac{1}{2}.$ (3) $-\infty.$ (5) $\frac{1}{2}.$ (6) $\sec^2 a.$

(8) Where $x=1$, parallel to x-axis; where $x=-2$, slope 9; where $x=0$, slope -3.

(9) 7 lb. per sq. in. per second. (10) $-\tan\dfrac{x}{2}$; $-\frac{1}{2}\sec^2\dfrac{x}{2}.$

(13) $y=\frac{2}{3}x^{\frac{3}{2}}+\frac{2}{3}.$ (14) $18(x-\sin x)^2.$ (15) $-\dfrac{\sin^{-1}\sqrt{x}}{2\sqrt{(1-x)}}.$

(16) $2\sec^3 x.$ (17) $-\dfrac{e^y+e^{-y}}{e^y-e^{-y}}.$ (20) $\dfrac{h^2-ab}{(hx+by)^3}.$

(30) Chord XY subtends a right angle at centre of first circle.

(33) $\dfrac{a^2+8b^2}{8b}$; $a-b.$ (35) $\theta=60°$; $\theta=\tan^{-1}\dfrac{2a}{b}.$

(39) $x(3m+m^3)-2y=am^3$; $2x+y(3m+m^3)=am^2(2+m^2).$

(40) $x\sin^2\theta+y\cos^2\theta=a\sin^2\theta\cos^2\theta$;
$x\cos^2\theta-y\sin^2\theta=a(\cos^6\theta-\sin^6\theta).$

(42) The points are determined by $\tan 2t=\pm 2.$

(46) $\dfrac{\pi}{3}.$ (48) $\left(\dfrac{a}{5},\ -\dfrac{a}{10}\right).$ (49) $2\sec x$, $2\sec x\tan x.$

(50) $\dfrac{2xy}{(x-3y^2)^3}.$ (51) $\dfrac{\sin(x+y)}{\cos^3(x+y)}.$ (52) $\frac{1}{2}\cos^3 2t.$

(55) $\tan^2\theta\,\dfrac{d^2y}{d\theta^2}-(\tan^3\theta-\tan\theta)\dfrac{dy}{d\theta}+y=0.$ (64) $-\frac{1}{4}.$

(65) $-\frac{1}{2}$. (66) $e^{x\cos a}\cos(x\sin a+a)$. (67) $\dfrac{b\sqrt{s^2-a^2}}{s}$.

(71) If one rod makes an angle a with the horizontal, the sliding rod makes an angle a with the other.

(77) $2\sqrt{2}a$. (80) $\left[\dfrac{c}{2}\left(3t+\dfrac{1}{t^3}\right),\ \dfrac{c}{2}\left(\dfrac{3}{t}+t^3\right)\right]$.

(81) $\left(\dfrac{3at}{2},\ -\dfrac{at^3}{2}\right)$. (83) $\left[12at^3+4at,\ -\dfrac{27at^4+6at^2}{2}\right]$.

(86) $2xy=\pm c^2$. (87) $(ax)^{\frac{2}{3}}+(by)^{\frac{2}{3}}=(a^2-b^2)^{\frac{2}{3}}$.

(90) $x-y=\dfrac{2a}{3}$; $x-y=a$, at $(0,\ -a)$;

$x-y=-\dfrac{a}{3}$, at $\left(-\dfrac{2a}{9},\ \dfrac{a}{9}\right)$.

(91) $x-y=2$ or $-\frac{1}{2}$; $x+y=\frac{1}{2}$ or -2.

(93) $xy(x+2y)-4xy+4x+3y-4=0$.

(94) About ·54 square foot.

(99) Hint: If x_1 is a root of $ax^2+2bx+c=0$, $f(x_1)$ may be reduced to a linear expression in x_1.

INDEX

The numbers refer to pages.

The numbers refer to pages.

The numbers refer to pages.

PART II

ELEMENTARY
INTEGRAL CALCULUS

PREFACE

THE object of this little work is to provide an introduction to the subject of Integral Calculus for mathematical and scientific students. The plan of the book is largely based on the Syllabus for the Higher Certificate Examination and it owes its origin to a set of lectures given by the author in preparation for this examination. The author ventures to hope that, while reducing the ground to be covered, he has not left out anything which is fundamentally necessary to the opening stages of the subject. A previous knowledge of the differentiation of powers, products, sines, cosines, etc. is however assumed.

No attempt has been made to treat the subject from an absolutely mathematical standpoint. The author fails to see the value of discussing the validity of a process until the nature of the process and its usual applications are thoroughly grasped. In fact, the work is not by any means intended as the final reading for a prospective mathematical scholar.

It is the author's opinion that, in many books, sufficient stress is not laid on what he has ventured to call the Fundamental Theorem. Historically the discovery of this theorem by Newton and Leibnitz was the turning point of the whole theory. He has therefore reversed the conventional order of treatment of the subject, and introduced the notion of summation only after the student has learnt how to integrate. It is hoped that this will lead to a deeper understanding of the Fundamental Theorem, and hence to a more skilful use of it.

In view of the recent regulations permitting the Preliminary Examination in Natural Science at Oxford to be taken from school, a chapter on Differential Equations has been added, making the book a complete preparation for the Integral required. The addition of questions on rotational dynamics to the Mechanics Syllabus of the Higher Certificate Examination seems to indicate a move in this direction, whilst from the point of view of general education the popularity of wireless renders some mathematical knowledge on the subjects of "damping" and "resonance" highly desirable.

A short historical survey has been added, for the author believes that this is a subject which might, with advantage, be taught to a greater extent than is done to-day.

His acknowledgements are due to Mr T. W. Chaundy, M.A., Tutor of Christ Church, Oxford, who has, in the press of much more important work, been good enough to read the MS. and has made many valuable suggestions. The extent to which the author is indebted to many standard works will be apparent to any mathematician who reads this volume. He desires to mention in particular the works of Williamson, Gibson and Edwards.

Questions from the Higher Certificate Examination are included by kind permission of the Oxford and Cambridge Schools Examination Board, and are marked in the text with an asterisk.

In conclusion the author wishes to record his indebtedness to the readers of the Cambridge University Press, to whose careful and excellent work the elimination of many errors is due.

G. L. P.

November, 1925.

CONTENTS

CHAPTER V

TRIGONOMETRICAL INTEGRALS AND FORMULÆ OF REDUCTION

CHAPTER VI

THE FUNDAMENTAL "INVERSION THEOREM." APPLICATION OF CALCULUS TO FINDING OF AREAS

CHAPTER VII

FURTHER APPLICATIONS TO GEOMETRY: SECTORIAL AREAS, VOLUMES OF REVOLUTION, FINDING THE LENGTH OF A CURVE

CHAPTER VIII

APPLICATION TO PROBLEM OF FINDING CENTRES OF GRAVITY

CHAPTER IX

FURTHER APPLICATIONS TO MECHANICS. RIGID DYNAMICS

CHAPTER X

DIFFERENTIAL EQUATIONS. A FEW TYPES

HISTORICAL SKETCH

THE arrangement of the subject-matter of Integral Calculus adopted in the text is almost exactly opposite to the order of the historical development of the subject. A little reflection will convince the student that this is the case in most branches of science. Attention is first drawn to some particular problem, insoluble by known methods. A method is discovered for this particular problem, and is gradually proved to be true for a large class of allied problems. Finally a series of rules governing the use of the method is drawn up. In teaching the subject, of course, the rules have to be learnt first and the application follows. This is exactly the history of the Integral Calculus.

It was quite inevitable that the attention of mathematicians should be drawn, sooner or later, to the problem of summation. As early as the third century B.C. we find the great geometer and physicist Archimedes determining the area bounded by a parabola, by splitting it up into rectangles and finding the sum of these rectangles. He arrived at the correct result, which shows that he knew the result, which we should express in the form

$$\int_0^1 x^3\, dx = \left[\frac{x^4}{4}\right]_0^1,$$

though, of course, he had not this notation. Very little advance was made for many years. The Hindu and Arab mathematicians who succeeded the Greek school concentrated their attention mainly on algebra and trigonometry and paid no great attention to questions of area and so forth. They were, however, responsible for the formulæ

$$\sum_{1}^{n} r^2 = \frac{n(n+1)(2n+1)}{6},$$

$$\sum_{1}^{n} r^3 = \frac{n^2(n+1)^2}{4},$$

which were of course used by later writers in performing summations.

But it is not till the time of Johann Kepler and Bonaventura Cavalieri that we find any definite traces of a general attempt to attack the problem of summation. In 1609 Kepler published his work "On the motion of the planet Mars" and it is clear that he was in possession of some method for finding the areas of focal sectors of an ellipse, whilst a few years later in another work he regards the volume of a cask as composed of thin circular discs, just as we should. Cavalieri, who was a Jesuit, published in 1635 the "Method of Indivisibles." This method, though his statement of it was not mathematically rigid, was in effect the method of elements that we use to-day. He stated, for instance, that a line is composed of a number of points without magnitude, a plane area of a number of lines without breadth, and a volume of a number of planes without thickness. He also arrived at the result

$$\int_0^a \frac{b^n x^n}{a^n} \, dx = \frac{ab^n}{n+1},$$

which is equivalent to the rule for integrating x^n.

The method of indivisibles found many exponents, and in the mathematical circles of correspondence* of which P. de Carcany and M. Mersenne were the intermediaries,

---

* These circles were the predecessors of the scientific societies of to-day, and were the normal means of spreading results at this period.

many definite integrations were exchanged. Notably we may mention Pascal (1623–62), who used the method in 1659 to find the area, volume of revolution and centre of gravity of the cycloid, and Wallis (1616–1703), a Cambridge mathematician who migrated to Oxford as Savilian Professor of Geometry. He expressed a complete set of rules for the summation of the series $\Sigma x^n dx$ for all cases except $n = -1$. His celebrated formulæ mentioned in Ch. V were obtained by regarding the area of a circle $\{y = (x - x^2)^{\frac{1}{2}}\}$ as intermediate between that of the curves $y = (x - x^2)^0$, $y = (x - x^2)^1$. A few years later he also gave some formulæ for the rectification of plane curves. Many other mathematicians of this period effected summations of this type, amongst whom we may mention Fermat, Edward Wright (who had a formula equivalent to $\int \sec \theta d\theta$), Roberval, Huyghens, and Torricelli.

We thus see that up to 1660 mathematicians had a good idea of the far-reaching results that might be effected by a general method of summation, and could perform many individual summations, but had no general rule for performing them. The discovery of the Theory of Fluxions by Newton and the Differential Calculus by Leibnitz at about the same date immediately linked up these results in the manner described in Ch. VI on the Fundamental Theorem. Almost immediately, therefore, a complete Integral Calculus sprang into being; the dry bones had long been there, this was the breath that gave them life. Space prohibits us from going into the long and bitter controversy which raged for over 100 years around the two principal figures in these discoveries, and which in some respects had a most unfortunate effect on English mathematics. Suffice it, however, to say that Newton had used his methods in the tract "de Quadratura Curvarum" many years before they were published—and that he knew the outline of the work at any rate as early as 1666. We do know definitely that Leibnitz saw some MSS of Newton's, during his visit to London in 1673, and quite possibly again in 1675 on a visit to Tschirnhausen, but we have no very clear evidence as to what ideas he derived from them. It is only fair to say that the controversy was not sought by the principal figures but by some of their followers, and it is unpleasant to have to add that many of Newton's supporters were more distinguished by their zeal than by their discretion or skill. Only one advantage arose out of this most unfortunate episode; I refer to the "challenge problems" sent out by both sides —at first merely as tests in the efficacy of their own methods, and later as annoyances to the other side. As these problems were often solved by both parties, they contributed but little to the controversy. They have however contributed many results, notably in Dynamics and Kinematics, which remain in the text-books to this day.

There is the other sad side to this subject. The main result was a standing aside of the English mathematicians from the results achieved on the Continent and a persistence in using the old notation and methods, which held English mathematics back at least 100 years, and from which we have hardly even now recovered. It was actually not till 1819 that Peacock, Herschel, and Babbage (the former of whom was moderator that year) were able to introduce the continental notation—or as they themselves cleverly put it* "to advocate the principles of pure

* In the Newtonian notation $\dot{x}$ is used where the continental writes $\dfrac{dx}{dt}$.

'd'-ism instead of the 'dot'-age of the University." So that finally the nett result was a grafting of the Leibnitzian notation on to Newtonian ideas.

However, four of Newton's contemporaries and immediate successors made some contributions to the theory. Brook Taylor (1685–1731) introduced the idea of changing the variable (Ch. III). Partial Fractions were studied by Cotes (1682–1716), whose early death robbed England of a second Newton, and his work was completed by de Moivre ; whilst a standard treatise on the whole method was written by Maclaurin in 1742.

But in the main the development of the subject on the Continent was far in advance of that in this country, owing to the superiority of the notation employed. Leibnitz himself had invented the symbols $\int$ (at first he wrote " omnia y " instead of $\int y$) and $\dfrac{d}{dx}$, and the methods were enthusiastically supported by the brothers Bernoulli (James and John) who were successively professors at Bâle. James Bernoulli wrote in 1691 the first treatise on Integral Calculus, Count J. F. Riccati spread the new doctrines in Italy, and d'Alembert (1717–83) in his " Traité de dynamique" showed how the principles governing the motion of a body could be expressed by the new notation. The subject of Differential Equations naturally grew up at the same time, but its development has taken much longer and in fact is not yet finally completed. The theory of the simple equations given in this book is however mainly due to d'Alembert and Leonhard Euler (1707–83). The latter made so many valuable contributions to all branches of mathematical research that it is hardly possible to find any branch of the subject which has not a theorem bearing his name associated with it. He published a vast quantity of works, some of which are still read, including (in 1770) his "Institutiones Calculi Integralis," a very full account of the theory, including the method of successive reductions and the Beta and Gamma Functions. The properties of Resonance, which are mentioned in the last chapter, were studied by Lagrange, who published in 1772–85 a series of tracts on Differential Equations which practically forms the basis of the whole subject as read to-day.

The student who is desirous of further historical information is referred to the standard works, particularly those by the late W. W. R. Ball and Prof. F. Cajori, to which the author himself is mainly indebted. It is the author's opinion that far too little is learnt about those mighty men who laid the foundations of our knowledge, and that, if we studied their history more, we might catch perchance a little of the great enthusiasm and inspiration with which they made their immortal discoveries.

CHAPTER I

DEFINITIONS AND STANDARD FORMS

1. Definition of Integration.

Integration is defined as the reversal of the process of differentiating. In the simpler cases one variable only is employed. Hence the process of "integrating" a given function of x is equivalent to that of finding a second function of x, which is such that the first function is the "differential" or "derivative" of the second. Or to put the matter into mathematical language—to integrate $F(x)$ is to find another function $\phi(x)$, such that $\dfrac{d}{dx}\phi(x) = F(x)$.

There is a standard notation for this, which is now almost universally adopted, viz. $\int F(x)\,dx$—which should be read "the integral of $F(x)$ with respect to x." The symbol $\int$ is in reality a conventionalized capital S and signifies the fact that integration is connected with the process of summation, as we shall see in Chapter VI.

2. Character of Integration.

A consideration which strikes us at the outset, is that in the final resort integration is a species of guesswork, i.e. we have to be able to see by intuition the kind of result which will, when differentiated, produce the function given. But it is not a process of illogical or unreasoning guesswork. The reader will find that there are definite things to try and definite rules to follow and to the total collection of these rules is given the name of "Integral Calculus."

We also remember that the differential of a constant is

zero. It follows, then, that when we try to reverse the process of differentiating we do not know whether to add a constant to the result or not, or, in any case, what constant to add.

For instance $\frac{d}{dx}(x^4)$, $\frac{d}{dx}(x^4 + 4)$ and $\frac{d}{dx}(x^4 + 100)$ are all equal to $4x^3$. Which of these functions then is the "integral" of $4x^3$? We get over the difficulty by saying that $\int 4x^3\,dx$ is $x^4 + C$, where C is a constant that remains to be determined or, as we usually put it, "an arbitrary constant." In any practical question, sufficient data are usually given to determine the value of this constant.

3. Indefinite Integrals. Necessity for Inclusion of Constant.

When an integral is given in this form it is known as an "indefinite" integral. The arbitrary constant should always be included unless it is quite clearly non-existent; its omission can often lead to serious error, as the following example will show.

We remember from Differential Calculus that

$$\frac{d}{dx}(-\cos^{-1}x) \text{ and } \frac{d}{dx}(\sin^{-1}x) \text{ are both equal to } \frac{1}{\sqrt{1-x^2}}.$$

It follows then (if we omit the arbitrary constants) that $\sin^{-1}x$ and $-\cos^{-1}x$ are equal, or that $\sin^{-1}x + \cos^{-1}x = 0$, a result which is plainly ridiculous.

What we should have written is

$$\text{(i)} \quad \int \frac{1}{\sqrt{1-x^2}} = \sin^{-1}x + C,$$

or $\qquad$ (ii) $\quad \int \frac{1}{\sqrt{1-x^2}} = -\cos^{-1}x + D;$

whence $\sin^{-1}x + C = -\cos^{-1}x + D,$

or $\sin^{-1}x + \cos^{-1}x = D - C.$

This result is not impossible since we know that

$$\sin^{-1}x + \cos^{-1}x = \frac{\pi}{2},$$

and we have merely to choose the constants D and C so that they differ by $\frac{\pi}{2}$.

4. Since, as we have seen, integration and differentiation are reverse processes, it follows that the integral of the sum of a number of separate functions is the sum of the integrals of the separate functions,

viz. $\int (u+v+w+\ldots)dx = \int u\,dx + \int v\,dx + \int w\,dx + \ldots.$

It is also obvious that integration is commutative with respect to a constant multiplier,

i.e. if A is constant $\int A u\,dx = A \int u\,dx.$

A table of Standard Forms can be constructed by reversing well-known results of Differential Calculus. Such a table is given here but the reader will find it of far greater value to construct his own table and keep it for reference, adding to it from time to time any new results he may meet. The results are given here both in the Differential and Integral Forms but in all cases the arbitrary constant has, for considerations of space, been left out. It is, of course, much more essential that such a table should exist in the reader's memory than in the text-book, as a thorough knowledge of it is essential to his further progress.

5. Table of Standard Forms.

Differential Calculus	*Integral Calculus*
$\dfrac{d}{dx}x^n = nx^{n-1}.$	$\displaystyle\int x^n\,dx = \dfrac{x^{n+1}}{n+1}$ unless $n=-1$.
$\dfrac{d}{dx}(ax+b)^n = na(ax+b)^{n-1}.$	$\displaystyle\int (ax+b)^n\,dx = \dfrac{(ax+b)^{n+1}}{(n+1)a}.$
$\dfrac{d}{dx}\log_e x = \dfrac{1}{x}.$	$\displaystyle\int \dfrac{1}{x}\,dx = \log_e x.$
$\dfrac{d}{dx}e^{ax} = ae^{ax}.$	$\displaystyle\int e^{ax}\,dx = \dfrac{1}{a}e^{ax}.$
$\dfrac{d}{dx}\sin x = \cos x.$	$\displaystyle\int \cos x\,dx = \sin x.$
$\dfrac{d}{dx}\cos x = -\sin x.$	$\displaystyle\int \sin x\,dx = -\cos x.$
$\dfrac{d}{dx}\sin mx = m\cos mx.$	$\displaystyle\int \cos mx\,dx = \dfrac{1}{m}\sin mx.$
$\dfrac{d}{dx}\tan x = \sec^2 x.$	$\displaystyle\int \sec^2 x\,dx = \tan x.$
$\dfrac{d}{dx}\cot x = -\operatorname{cosec}^2 x.$	$\displaystyle\int \operatorname{cosec}^2 x\,dx = -\cot x.$
$\left.\begin{array}{l}\dfrac{d}{dx}\sin^{-1}\dfrac{x}{a} = \dfrac{1}{\sqrt{a^2-x^2}}\\[2mm]\dfrac{d}{dx}\cos^{-1}\dfrac{x}{a} = -\dfrac{1}{\sqrt{a^2-x^2}}\end{array}\right\}.$	$\displaystyle\int \dfrac{dx}{\sqrt{a^2-x^2}} = \left\{\begin{array}{l}\sin^{-1}\dfrac{x}{a},\\[2mm]-\cos^{-1}\dfrac{x}{a}.\end{array}\right.$
$\dfrac{d}{dx}\tan^{-1}\dfrac{x}{a} = \dfrac{a}{x^2+a^2}.$	$\displaystyle\int \dfrac{dx}{a^2+x^2} = \dfrac{1}{a}\tan^{-1}\dfrac{x}{a}$ $\left(\text{or } -\dfrac{1}{a}\cot^{-1}\dfrac{x}{a}\right).$
$\dfrac{d}{dx}a^x = a^x\log_e a.$	$\displaystyle\int a^x\,dx = \dfrac{a^x}{\log_e a}.$

N.B. (1) Notice that all integrals commencing with the prefix "co-" have a negative sign prefixed.

(2) In cases like $\displaystyle\int \dfrac{dx}{a^2+x^2}$ and $\displaystyle\int \dfrac{dx}{\sqrt{a^2-x^2}}$ it is useful to notice the

dimensions: for instance $\int \dfrac{dx}{a^2+x^2}$ is of dimensions -1, but $\tan^{-1}\dfrac{x}{a}$ is of zero dimensions, hence we clearly must put in a factor $\dfrac{1}{a}$ of dimensions -1; on the other hand $\int \dfrac{dx}{\sqrt{a^2-x^2}}$ and $\sin^{-1}\dfrac{x}{a}$ are both of zero dimensions and no factor is required.

(3) It is interesting to notice that the case $\int \dfrac{1}{x}\,dx$ is not really an exception to the rule for $\int x^n dx$.

For
$$\int x^n dx = \frac{x^{n+1}}{n+1} + A,$$
or as we may equally put it
$$= \frac{x^{n+1}-1}{n+1} + B,$$
where B is another arbitrary constant.

Now consider the Lt as $n \to -1$.

Then
$$\int x^{-1} dx = \underset{n \to -1}{\text{Lt}}\ \frac{x^{n+1}-1}{n+1} + B$$
$$= \underset{m \to 0}{\text{Lt}}\ \frac{x^m-1}{m} + B,$$
putting $m = n+1$.

But by a Theorem on Limits (Chrystal's *Algebra*, ch. xxv, § 13)
$$\underset{m \to 0}{\text{Lt}}\ \frac{x^m-1}{m} = \log x.$$
$$\therefore \int x^{-1} dx = \log x + B.$$

EXAMPLES I A

Integrate the following:

(1) $\sqrt{x},\quad \dfrac{1}{\sqrt{x}},\quad \dfrac{1+x+x^2}{3}$.

(2) $\sqrt[3]{x^4},\quad \dfrac{1}{x^6},\quad \dfrac{1}{\sqrt[5]{x^7}}$.

(3) $\sqrt[3]{3x-4},\quad \sqrt{3x-4},\quad \dfrac{1}{\sqrt{3x-4}}$.

(4) $\dfrac{1}{a+x},\quad \dfrac{1}{a+x}+\dfrac{1}{a-x},\quad \dfrac{1}{(a-x)^n}$.

(5) $\dfrac{1}{x+1}$, $\dfrac{x+3}{x+1}$, $\dfrac{1}{ax+b}$, $\dfrac{ax+c}{ax+b}$, $\dfrac{Ax+B}{ax+b}$.

(6) $\dfrac{1+x}{\sqrt{x}}$, $\dfrac{x^2+1}{x}$, $\dfrac{1}{\sqrt{x+1}+\sqrt{x}}$, $x\sqrt{1+x}$.

(7) Verify by differential that $\displaystyle\int \dfrac{dx}{\sqrt{x^2+a^2}}=\log(x+\sqrt{x^2+a^2})+c$.

(8) Find by trial, noticing result of Ex. 7, $\displaystyle\int \dfrac{dx}{\sqrt{x^2-a^2}}$.

6. Two Special Functional Forms.

Remembering that $\dfrac{d}{dx}\log\{\phi(x)\}=\dfrac{\phi'(x)}{\phi(x)}$, it is clear that

$$\int \frac{\phi'(x)\,dx}{\phi(x)}=\log\phi(x)+C.$$

Or putting the matter into words: "the integral of a fraction whose numerator is the differential of its denominator is the logarithm of the denominator." We shall have occasion to use this form in Ch. IV, and the rule is of frequent general application.

We see also that

$$\int [\phi(x)]^n \cdot \phi'(x)\,dx=\frac{1}{n+1}[\phi(x)]^{n+1}+C.$$

A few examples of the applications of these rules are given:

(1) $\displaystyle\int \cot x\,dx=\int \dfrac{\cos x}{\sin x}\,dx$.

    Here $\phi(x)=\sin x$.

    Hence $\displaystyle\int \cot x\,dx=\log\sin x+C$.

    Again

$$\int \tan x=-\log\cos x+C \text{ or } \log\sec x+C.$$

(2) Similarly $\displaystyle\int \dfrac{2ax+b}{ax^2+bx+c}\,dx=\log(ax^2+bx+c)+C$.

(3) $\int \dfrac{\cos x}{\sin^2 x}\, dx = \int (\sin x)^{-2} . \cos x\, dx.$

Here $\phi(x) = \sin x.$

$$\therefore \int \dfrac{\cos x}{\sin^2 x}\, dx = -\dfrac{1}{\sin x} = -\operatorname{cosec} x.$$

Or again $\int \sin^4 x \cos x\, dx = \dfrac{\sin^5 x}{5} + C.$

Lastly to find $\int \tan^3 x\, dx.$

$$\int \tan^3 x\, dx = \int \tan x . \tan^2 x\, dx$$

$$= \int \tan x\, (\sec^2 x - 1)\, dx$$

$$= \int \tan x \sec^2 x\, dx - \int \tan x\, dx$$

$$= \frac{1}{2} \tan^2 x + \log \cos x\,;$$

this is seen to be an example of the use of both rules.

The student should observe that these devices are merely methods of representing the reversal of the rule for differentiating a "function of a function"; and if the student has become skilled in differentiating expressions of this sort he will have little difficulty in integrating, or in recognising how a given expression might be derived.

In some of the examples following, it may be necessary to employ some transformation before commencing to integrate. This is often the case with trigonometrical functions such as $\cos^2 x$, etc., which should always be expressed in terms of double the angle.

EXAMPLES I B

Integrate the following:

(1) $2 \sin x \cos x$, $\cos^2 x$, $\cos mx \cos nx$, $\sin mx \sin nx$.

(2) $\tan x$, $\dfrac{\sin x}{\cos^2 x}$, $\cos x \left\{ \dfrac{1}{\sin x} + \dfrac{1}{\sin^3 x} \right\}$, $\dfrac{\sec^2 x}{\tan x}$.

(3) $\dfrac{1}{x \log x}$, $\dfrac{e^x}{e^x + 2}$, $\dfrac{e^x + e^{-x}}{e^x - e^{-x}}$, $\dfrac{\cot x}{\log \sin x}$.

(4) $\dfrac{2x - 3}{x^2 - 3x + 2}$, $\dfrac{2x + 1}{2x - 1}$, $(x^2 + 2x + 2)(x + 1)$, $\dfrac{x^2 + 2x + 2}{x + 1}$, $\dfrac{x + 1}{x^2 + 2x + 2}$.

(5) $\dfrac{1}{x^2 + 9}$, $\dfrac{x^2}{x^2 + 4}$, $\dfrac{2x}{1 + x^4}$, $\dfrac{1}{7 + 3x^2}$.

(6) $\cos^3 x \sin x$, $\sin^2 x \cos^3 x$, $\cos^3 x$.

CHAPTER II

INTEGRATION BY SUBSTITUTION,
OR
CHANGING THE SUBJECT OF INTEGRATION

7. Definite Integrals.

We have already spoken of indefinite integrals in the last chapter. Before proceeding further we ought to define what is meant by a "definite integral."

The expression $\int_a^b f(x)\,dx$ is called "the definite integral of $f(x)$ between the limits of integration b and a." Its value is formed by first calculating the indefinite integral of $f(x)$ and then subtracting the two values of this expression when $x = b$ and when $x = a$, from one another, the value for the upper limit being always taken first.

For example, to find $\int_a^b x^2\,dx$,

$$\int x^2\,dx = \frac{x^3}{3} + C,$$

$$\therefore \text{ definite integral } \int_a^b x^2\,dx = \left[\frac{x^3}{3} + C\right]_a^b$$

$$= \frac{b^3}{3} + C - \frac{a^3}{3} - C$$

$$= \frac{b^3 - a^3}{3}.$$

We notice that in a definite integral the arbitrary constant always disappears, and consequently it is usually omitted.

We shall see later in Ch. VI that a physical meaning can often be assigned to a Definite Integral, and it is this type of integral which makes its appearance in questions connected with Geometry, Mechanics, etc.

8. Changing the subject of Integration.

As we have noticed, Integration is, in the last resort, a species of guesswork. It very often happens, however, that the function to be integrated is not expressed in a form which allows us to see readily of what function it is the differential. A large part of the theoretical side of the Integral Calculus consists of various devices by which we can reduce an apparently difficult integral to a form in which it can be more readily recognized.

The first and most important of these processes is that of Substitution or "Changing the subject of Integration" and it is very important that the student should master this thoroughly before going any further.

Suppose that we have an integral, $\int V\,dx$, which appears to be insoluble as it stands, but that we can see that the form of the expression V can be simplified if, instead of some function of x, we write some other variable z (such as, for instance, $z = \cos x$, or $z = e^x$).

Now let $u = \int V\,dx$.

Suppose that when we put z for x the expression V changes into a new expression V_z.

Now by the definition of an integral

$$\frac{du}{dx} = V.$$

But since z and x are connected

$$\frac{du}{dx} = \frac{du}{dz} \cdot \frac{dz}{dx}.$$

$$\therefore \frac{du}{dz} = V \frac{dx}{dz}.$$

The right-hand side of this equation can now be expressed entirely in terms of z and we get

$$\frac{du}{dz} = V_z \frac{dx}{dz}.$$

$$\therefore u = \int V_z \frac{dx}{dz} \cdot dz.$$

We thus see that we can, without altering the result, change the variable from x to z provided we insert instead of dx the expression $\dfrac{dx}{dz}\,dz$. When the result is written entirely in terms of z, it may be easier to integrate than the original expression in terms of x.

9. If the original integral is indefinite, the result must be converted back so that it appears in terms of x when the integration is completed. If however the integral is a definite integral, we can avoid this process, which is often laborious, by changing the limits. We take as the new limits the value of z corresponding to the limiting values of x, and the definite integral can then be found without reverting to the original variable. The choice of the right substitution often demands considerable ingenuity, and the whole method is best illustrated by worked examples, which we give in the following section.

Many new standard forms can be found in this way, and the student is advised to add the results printed in heavier type to his table of Standard Forms.

10. Worked Examples.

(i) We first choose an example of which we know the result,

$$\text{viz.} \int \frac{1}{\sqrt{1-x^2}}\,dx.$$

It is seen that if we put $x = \sin\theta$ the radical disappears, giving $\cos\theta$ in the denominator.

The work is usually arranged as follows:

Let $x = \sin\theta$; then $\quad \dfrac{dx}{d\theta} = \cos\theta,$

i.e. for dx we put $\cos\theta\,.\,d\theta.$

$$I = \int \frac{1}{\sqrt{1-\sin^2\theta}}\,.\,\cos\theta\,.\,d\theta$$

$$= \int \frac{1}{\cos\theta}\,.\,\cos\theta\,.\,d\theta$$

$$= \int 1\,d\theta = \theta + C.$$

But $\theta = \sin^{-1}x.$

Hence $\qquad \displaystyle\int \frac{dx}{\sqrt{1-x^2}} = \sin^{-1}x + C.$

If we had chosen the substitution $x = \cos\theta$ we should have arrived at the other form of the integral which has already been noted in Ch. I, viz. $-\cos^{-1}x + D.$

(ii) $\displaystyle\int \sqrt{a^2 - x^2}\,dx.$

Again put $\quad x = a\sin\theta$, then $\dfrac{dx}{d\theta} = a\cos\theta$

or $\qquad\qquad\qquad\qquad\qquad dx = a\cos\theta\,.\,d\theta.$

$$I = \int \sqrt{a^2 - a^2 \sin^2 \theta} \,.\, a \cos \theta \,.\, d\theta$$

$$= \int a^2 \cos^2 \theta \,.\, d\theta$$

$$= a^2 \int \frac{1 + \cos 2\theta}{2} \, d\theta$$

$$= \frac{a^2}{2} \int 1 \, d\theta + \frac{a^2}{2} \int \cos 2\theta \, d\theta$$

$$= \frac{a^2 \theta}{2} + \frac{a^2 \sin 2\theta}{4}$$

$$= \frac{a^2}{2} [\theta + \sin \theta \cos \theta]$$

$$= \frac{a^2}{2} \left[\sin^{-1} \frac{x}{a} + \frac{x}{a} \sqrt{1 - \frac{x^2}{a^2}} \right]$$

$$= \frac{a^2}{2} \sin^{-1} \frac{x}{a} + \frac{x \sqrt{a^2 - x^2}}{2}.$$

Hence $\quad \int \sqrt{a^2 - x^2} \, dx = \dfrac{a^2}{2} \sin^{-1} \dfrac{x}{a} + \dfrac{x \sqrt{a^2 - x^2}}{2}.$

(iii) Integrals containing the expression $(x^2 + a^2)$ can often be simplified by putting $x = a \tan \theta$.

Let us try $\qquad \displaystyle\int_0^a \frac{x}{\sqrt{x^2 + a^2}} \, dx.$

Let $x = a \tan \theta$, i.e. $dx = a \sec^2 \theta \, d\theta$.

Also when $x = a$, $\theta = \dfrac{\pi}{4}$; when $x = 0$, $\theta = 0$.

Hence the new limits are $\dfrac{\pi}{4}$ and 0.

$$\therefore I = \int_0^{\frac{\pi}{4}} \frac{a\tan\theta}{\sqrt{a^2 + a^2\tan^2\theta}} . a\sec^2\theta \, d\theta$$

$$= \int_0^{\frac{\pi}{4}} a\tan\theta\sec\theta \, d\theta$$

(remembering $\sec^2\theta = 1 + \tan^2\theta$)

$$= a\int_0^{\frac{\pi}{4}} \frac{\sin\theta}{\cos^2\theta} \, d\theta$$

$$= a\left[\frac{1}{\cos\theta}\right]_0^{\frac{\pi}{4}} = a\,(\sqrt{2} - 1).$$

(iv) Algebraic substitutions are sometimes useful.

Consider $\qquad \int \dfrac{dx}{2\sqrt{x}\,.\,(1+x)}.$

Put $\sqrt{x} = z$; i.e. $\dfrac{1}{2\sqrt{x}}\,dx = dz$,

$$I = \int \frac{dz}{1 + z^2}$$

$$= \tan^{-1} z + C$$

$$= \tan^{-1}\sqrt{x} + C.$$

Note. A few further suggestions are appended for integrals of a more difficult character.

(1) Any expression containing a single linear surd, e.g. $\sqrt{ax+b}$, can be reduced to an algebraic form by the substitution $\sqrt{ax+b} = t$.

The integration may then be effected by the methods given and those of the following chapters.

(2) An expression containing a double surd of the form $\sqrt{(a-x)(x-b)}$ can be treated by the substitution

$x = a\cos^2\theta + b\sin^2\theta$, which renders it trigonometric and integrable as in Ch. V.

(3) Any expression containing a surd of the form $\sqrt{(x-a)(x-b)(x-c)}$ is not integrable in terms of known functions. Such an expression in fact gives rise to Elliptic Functions.

EXAMPLES II

Integrate by the method of substitution:

(1) $\sqrt{1-x}$.   (2) $\sqrt[3]{1+2x}$.   (3) $\dfrac{2x}{1+x^4}$.   (4) $\dfrac{3x^2}{1+x^6}$.

(5) $\dfrac{1}{x^2+2x+3}$ $\{z = x+1\}$.   (6) $\dfrac{1}{\sqrt{8-2x-x^2}}$.

(7) $\dfrac{1}{(x-1)\sqrt{x^2-2x}}$.   (8) $\dfrac{1}{\sqrt{x-x^2}}$ $\{z = x - \frac{1}{2}\}$.

(9) $\dfrac{1}{e^x+e^{-x}}$.   (10) $e^x\cos e^x$.   (11) Evaluate $\displaystyle\int_0^a \frac{x}{\sqrt{a^2-x^2}}\,dx$.

(12) Evaluate $\displaystyle\int_0^a x^2\sqrt{a^3+x^3}\,dx$.

(13) By writing $z = \tan\dfrac{x}{2}$, prove that

$$\int \operatorname{cosec} x\,dx = \log\tan\frac{x}{2} + C$$

and
$$\int \sec x\,dx = \log\tan\left(\frac{\pi}{4}+\frac{x}{2}\right) + C,$$

or
$$\log(\sec x + \tan x) + C.$$

(14) Use the last example to prove that

$$\int \frac{dx}{\sqrt{a^2+x^2}} = \log\frac{x+\sqrt{x^2+a^2}}{a} + C.$$

(15) Using result of Ex. 14, prove that

$$\int \sqrt{a^2+x^2}\,dx = \frac{x\sqrt{a^2+x^2}}{2} + \frac{a^2}{2}\log\left(\frac{x+\sqrt{x^2+a^2}}{a}\right) + C.$$

Find the following:

(16) $\int \dfrac{dx}{1+\cos x}$. (17) $\int \dfrac{dx}{x\sqrt{2ax-x^2}}$. (18) $\int \dfrac{dx}{x\sqrt{x^2-a^2}}$.

(19) $\int \dfrac{dx}{\sqrt{2ax-x^2}}$. *(20) $\int (\tan x + \tan^3 x)\,dx$.

*(21) $\int_0^{\frac{1}{\sqrt{2}}} x^2 \sqrt{1-x^2}\,dx$.

*(22) Show that $\int \dfrac{dx}{\sqrt{(x-a)(x-b)}}$ can be evaluated by the substitution $y^2 = \dfrac{x-a}{x-b}$.

*(23) Show that the integrals $\int \dfrac{1-x^2}{1+kx^2+x^4}\,dx$ and $\int \dfrac{1+x^2}{1+kx^2+x^4}\,dx$ can be found by the substitutions $v = x - \dfrac{1}{x}$, $u = x + \dfrac{1}{x}$.

*(24) By using the substitution $x+1 = \dfrac{1}{y}$, prove that

$$\int_1^3 \dfrac{dx}{(x+1)\sqrt{4x-3-x^2}} = \dfrac{\pi}{2\sqrt{2}}.$$

CHAPTER III

INTEGRATION BY PARTS

11. We have noticed already that the art of integration lies in transforming a difficult integral into one which can be recognized more readily as a known form. When the function to be integrated is in the form of a product, we can sometimes effect this by the device known as "Integration by Parts." This is the result in Integral Calculus which corresponds to the rule for differentiating a product and is easily found as follows.

We know that if u and w are functions of x, then

$$\frac{d}{dx}\{uw\} = w \cdot \frac{d}{dx}(u) + u \cdot \frac{d}{dx}(w).$$

Now if we integrate both sides of this equation we get

$$uw = \int w \cdot \frac{du}{dx} \cdot dx + \int u \cdot \frac{dw}{dx} \cdot dx \dots\dots\dots(A).$$

We can for the sake of convenience consider the arbitrary constants as included in the integrals.

Now u and w are perfectly general and unrestricted functions. Let us suppose that w is the integral of such other function v, i.e. that

$$w = \int v\,dx$$

or $\qquad \dfrac{dw}{dx} = v.$

Making this substitution in equation (A) we now have

$$u \cdot \int v\,dx = \int\left[\left\{\int v\,dx\right\} \cdot \frac{du}{dx}\right] dx + \int uv\,dx,$$

or rearranging

$$\int uv\,dx = u . \int v\,dx - \int \left(\int v\,dx \right) . \frac{du}{dx} . dx.$$

The result is perhaps more clearly stated in words as follows:

Integral of product = first function × Integral of second − Integral of {Differential of first × Integral of second}.

12. We notice, first of all, that the method does not "evaluate" the integral of a product; there is, in fact, no general formula to do this; what it does enable us to do is to connect the integral of a product with a second integral which may be easier than the original. The rule is, in fact, the simplest instance of a Reduction Formula, of which we shall have further examples in Ch. V. Care must, of course, be taken to choose the order of the functions in such a way as to reduce the second integral as far as possible. Otherwise we shall gain nothing by the transformation. In certain examples it may be necessary to apply the process more than once before we finally arrive at a result which can be integrated at sight. The necessity for a right choice of the order of the functions is illustrated by the following parallel attempts to evaluate the same integral:

$$\int x \cos x\,dx.$$

Let $u = x,$ $v = \cos x,$

$\dfrac{du}{dx} = 1,$ $\int v\,dx = \sin x.$

$I = x . \sin x - \displaystyle\int 1 . \sin x\,dx$

$= \mathbf{x\,sin\,x + cos\,x}.$

$$\int x \cos x\,dx.$$

Let $u = \cos x,$ $v = x,$

$\dfrac{du}{dx} = -\sin x,$ $\int v\,dx = \dfrac{x^2}{2}.$

$I = \dfrac{x^2}{2} \cos x + \displaystyle\int \dfrac{x^2}{2} \sin x\,dx,$

a result which is quite true but does not help us.

13. Worked Examples.

Careful note should be taken of the following worked examples:

(i) $\int x e^x \, dx.$ Let $u = x,$ $v = e^x,$

$$\frac{du}{dx} = 1, \qquad \int v \, dx = e^x$$

$$\therefore I = x \cdot e^x - \int 1 \cdot e^x \, dx$$

$$= x e^x - e^x + C$$

$$= (x - 1) e^x + C.$$

(ii) In some cases unity may be used as a function,

e.g. $\int \log x \, dx = \int 1 \cdot \log x \, dx.$ Let $u = \log x,$ $v = 1,$

$$\frac{du}{dx} = \frac{1}{x}, \qquad \int v \, dx = x.$$

$$\therefore I = x \log x - \int 1 \cdot dx$$

$$= x \log x - x + C$$

$$= x (\log x - 1) + C$$

or $$x \log \frac{x}{e} + C.$$

(iii) The process may have to be repeated two or more times before the evaluation can be effected.

Consider $\int x^2 \sin x \, dx.$ Here $u = x^2,$ $v = \sin x,$

$$\frac{du}{dx} = 2x, \quad \int v \, dx = -\cos x.$$

$$\therefore I = -x^2 \cos x + \int 2x \cos x \, dx.$$

To evaluate $\int x \cos x\, dx$ we repeat the process, getting the result $x \sin x + \cos x + C$ as in § 12.

Hence finally

$$I = -x^2 \cos x + 2x \sin x + 2 \cos x + C$$
$$= -\cos x\,(x^2 - 2) + 2x \sin x + C.$$

14. An Important Artifice.

We can sometimes by a little ingenuity bring the second integral back to the same form as the original integral, and thus find the value of the integral directly.

This is a very useful and important artifice and the example given should be carefully studied.

To find $\int \sqrt{a^2 + x^2}\, dx$. Let $u = \sqrt{a^2 + x^2}$, $v = 1$,

$$\frac{du}{dx} = \frac{x}{\sqrt{a^2 + x^2}}, \qquad \int v\, dx = x.$$

Then

$$I = x\sqrt{a^2 + x^2} - \int \frac{x^2}{\sqrt{a^2 + x^2}}\, dx$$

$$= x\sqrt{a^2 + x^2} - \int \left\{ \frac{x^2 + a^2}{\sqrt{a^2 + x^2}} - \frac{a^2}{\sqrt{a^2 + x^2}} \right\} dx$$

$$= x\sqrt{a^2 + x^2} - \int \sqrt{a^2 + x^2}\, dx + a^2 \int \frac{dx}{\sqrt{a^2 + x^2}}$$

$$= x\sqrt{a^2 + x^2} - I + a^2 \log \frac{x + \sqrt{x^2 + a^2}}{a} + C,$$

cf. Ex. II, 14.

$$\therefore\ 2I = x\sqrt{a^2 + x^2} + a^2 \log \frac{x + \sqrt{x^2 + a^2}}{a},$$

i.e. $\int \sqrt{a^2 + x^2}\, dx = \dfrac{x\sqrt{a^2 + x^2}}{2} + \dfrac{a^2}{2} \log \dfrac{x + \sqrt{a^2 + x^2}}{a} + C.$

This method is also applicable to

$$\int \sqrt{x^2 - a^2}\, dx, \quad \int \sqrt{a^2 - x^2}\, dx,$$

and the student should perform the process for these forms and make a careful note of the result.

15. The forms $e^{ax} \cos bx$, $e^{ax} \sin bx$.

These forms can at once be integrated by this method, two successive applications being necessary.

Let
$$I = \int e^{ax} \sin bx\, dx.$$

Integrate by parts, then

$$I = - e^{ax} \frac{\cos bx}{b} + \int \frac{a}{b} e^{ax} \cos bx\, dx.$$

But again

$$\int e^{ax} \cos bx\, dx = e^{ax} \frac{\sin bx}{b} - \int \frac{a}{b} e^{ax} \sin bx\, dx.$$

Therefore substituting

$$I = - e^{ax} \frac{\cos bx}{b} + \frac{a}{b^2} e^{ax} \sin bx - \frac{a^2}{b^2} I,$$

i.e.
$$\frac{a^2 + b^2}{b^2} I = \frac{a e^{ax} \sin bx - b e^{ax} \cos bx}{b^2},$$

or
$$I = \frac{e^{ax}}{a^2 + b^2} \{a \sin bx - b \cos bx\} + C$$

$$= \frac{e^{ax}}{\sqrt{a^2 + b^2}} \left\{ \frac{a}{\sqrt{a^2 + b^2}} \sin bx - \frac{b}{\sqrt{a^2 + b^2}} \cos bx \right\} + C.$$

Now let $\tan \alpha = \dfrac{b}{a}$, then

$$\frac{a}{\sqrt{a^2 + b^2}} = \cos \alpha \quad \text{and} \quad \frac{b}{\sqrt{a^2 + b^2}} = \sin \alpha.$$

Hence

$$I = \frac{e^{ax}}{\sqrt{a^2+b^2}}\{\sin bx \cos \alpha - \cos bx \sin \alpha\} + C$$

$$= \frac{e^{ax}}{\sqrt{a^2+b^2}} \sin (bx - \alpha) + C$$

$$= \frac{1}{\sqrt{a^2+b^2}} e^{ax} \sin \left(bx - \tan^{-1}\frac{b}{a}\right) + C*.$$

Similarly we can prove that

$$\int e^{ax} \cos bx\, dx = \frac{1}{\sqrt{a^2+b^2}} e^{ax} \cos \left(bx - \tan^{-1}\frac{b}{a}\right) + C'*.$$

Ex. Integrate $e^{3x} \sin 4x \cos 6x$.

$$I = \frac{1}{2}\int e^{3x} (\sin 10x - \sin 2x)\, dx,$$

$$2I = \int e^{3x} \sin 10x\, dx - \int e^{3x} \sin 2x\, dx$$

$$= \frac{e^{3x}}{\sqrt{109}} \sin \left(10x - \tan^{-1}\frac{10}{3}\right) - \frac{e^{3x}}{\sqrt{13}} \sin \left(2x - \tan^{-1}\frac{2}{3}\right) + C.$$

EXAMPLES III

Integrate the following :

(1) $xe^{-x}, \quad xe^{2x}, \quad x^2e^x, \quad x^3e^x$.

(2) $x \sin x, \quad x^2 \cos x, \quad x^3 \sin x, \quad x \cos^2 x, \quad x \sin x \cos x$.

(3) $x \log x, \quad x^2 \log x, \quad x^n \log x$.

(4) $e^x \cos x, \quad e^x \cos 2x, \quad e^x \sin 2x, \quad e^x (\sin x + \cos x)$.

(5) $\sin^{-1} x, \quad \tan^{-1} x, \quad x \sin^{-1} x, \quad x \tan^{-1} x$.

(6) $\sqrt{x^2-a^2}, \quad \sqrt{a^2-x^2}$.

(7) $e^{3x} \cos 4x, \quad 2e^{5x} \sin 7x \cos 3x, \quad 4e^x \sin x \sin 2x \sin 3x$.

(8) $x \sec^2 x, \quad \dfrac{\sin^{-1} x}{(1-x^2)^{\frac{3}{2}}}$.

---

* This can be directly deduced from the rule for differentiating these functions given in the standard books on Differential Calculus.

CHAPTER IV

RATIONAL ALGEBRAIC FRACTIONS

16. Reduction of Algebraic Fractions.

Frequently the integral to be found, either before or after the process of substitution, takes the form of a Rational Algebraic Fraction. As a preparatory step to the integration of such a form, the fraction must always be resolved into Partial Fractions. We assume that the student is already familiar with this process; if this is not the case he is referred to the standard text-books on Algebra.

We shall proceed to show that if this is done the result can practically always be expressed in terms of powers, logs, and inverse tangents.

If the numerator of the given fraction is of higher order than the denominator, we should first divide out as far as we can. The quotient thus formed can only consist of powers of x and constants and is immediately integrable.

In what follows we shall suppose this to have been done and concern ourselves only with the remaining fractional part, in which the numerator will be at least one lower in order than the denominator.

The four possible types of term.

There will, in general, be four types of term which can occur:

(1) Terms like $\dfrac{A}{Bx + C}$.

(2) Terms like $\dfrac{A}{(Bx + C)^n}$.

(3) Terms like $\dfrac{ax + b}{(Ax^2 + Bx + C)}$,

            where $Ax^2 + Bx + C$ has no real factors.

(4) Terms like $\dfrac{ax + b}{\{Ax^2 + Bx + C\}^n}$.

We shall confine our attention to the first three types, as the last, which involves the presence in the denominator of a doubly repeated irreducible quadratic factor, is uncommon and outside the scope of this work. The reader desirous of knowing how to deal with this case is referred to Williamson's *Integral Calculus*, p. 53, §§ 43, etc. We proceed to show that the other types are immediately integrable:

(1) $\displaystyle\int \frac{A\,dx}{Bx + C}$ is seen to be $\dfrac{A}{B} \log (Bx + C)$.

(2) $\displaystyle\int \frac{A\,dx}{(Bx + C)^n}$ is seen to be $- \dfrac{A}{(n - 1)\,B} \cdot \dfrac{1}{(Bx + C)^{n-1}}$.

Each of these may be verified by writing $Bx + C = Z$.

17. Integration of terms like $\dfrac{ax + b}{Ax^2 + Bx + C}$.

In order to integrate terms of type (3) *we first introduce into the numerator a multiple of the differential of the denominator sufficient to remove all the x's*, making the numerical factors right afterwards.

Thus for $\dfrac{ax + b}{Ax^2 + Bx + C}$ we put

$$\frac{a}{2A} \cdot \frac{2Ax + B}{Ax^2 + Bx + C} + \left(b - \frac{aB}{2A}\right) \frac{1}{Ax^2 + Bx + C}.$$

The integral of the term thus reduces to the sum of two integrals.

$$(a) \quad \frac{a}{2A} \int \frac{2Ax + B}{Ax^2 + Bx + C} dx,$$

and

$$(b) \quad \frac{2Ab - aB}{2A} \int \frac{dx}{Ax^2 + Bx + C}.$$

(a) is seen at once to be $\frac{a}{2A} \log (Ax^2 + 2Bx + C)$, ($\S 6$).

To deal with (b) we re-write the term as below

$$\frac{2Ab - aB}{2A} \int \frac{dx}{Ax^2 + Bx + C} = \frac{2Ab - aB}{2A^2} \int \frac{dx}{x^2 + \frac{B}{A} x + \frac{C}{A}},$$

$$= \frac{2Ab - aB}{2A^2} \int \frac{dx}{\left(x + \frac{B}{2A} \right)^2 + \frac{4AC - B^2}{4A^2}} dx.$$

Now since $Ax^2 + Bx + C$ has no real factors, $4AC - B^2$ is positive, i.e. the integral reduces to some integral of the form

$$K \int \frac{dx}{(x + M)^2 + N^2},$$

the result of which is seen to be $\frac{K}{N} \tan^{-1} \frac{x + M}{N}$.

$Ex.$ $\int \frac{x + 5}{x^2 + 4x + 7} dx = \frac{1}{2} \int \frac{2x + 4}{x^2 + 4x + 7} dx + \int \frac{3}{x^2 + 4x + 7} dx$

$$= \frac{1}{2} \log (x^2 + 4x + 7) + \int \frac{3}{(x + 2)^2 + 3} dx$$

$$= \frac{1}{2} \log (x^2 + 4x + 7) + \frac{3}{\sqrt{3}} \tan^{-1} \frac{x + 2}{\sqrt{3}} + C.$$

The reader should make no attempt to remember and quote these results as formulæ. It is far better to treat each example on its own merits. The articles above give the methods of attack and indicate the results likely to be found. The student with a knowledge of the use of imaginary quantities and de Moivre's Theorem will find that the case of irreducible factors can also be treated by the use of imaginary quantities, but this lies beyond the scope of the present volume.

18. Worked Examples.

(1) Find $\displaystyle\int \frac{dx}{x^2 - a^2}$.

Let $\displaystyle\frac{1}{x^2 - a^2} = \frac{A}{x - a} + \frac{B}{x + a}$.

Clearing of fractions we have $1 \equiv A(x + a) + B(x - a)$.

Since this is an identity we equate coefficients of powers of x.

$$\therefore \ A + B = 0,$$
$$A - B = \frac{1}{a},$$

whence $\qquad A = \dfrac{1}{2a}, \quad B = -\dfrac{1}{2a}.$

$$\therefore \ \frac{1}{x^2 - a^2} \equiv \frac{1}{2a} \cdot \frac{1}{x - a} - \frac{1}{2a} \cdot \frac{1}{x + a}.$$

$$\therefore I = \frac{1}{2a} \int \frac{1}{x - a}\, dx - \frac{1}{2a} \int \frac{1}{x + a}\, dx$$

$$= \frac{1}{2a} \log(x - a) - \frac{1}{2a} \log(x + a) + C$$

$$= \frac{1}{2a} \log \frac{x - a}{x + a} + C.$$

Hence $\qquad \displaystyle\int \frac{dx}{x^2 - a^2} = \frac{1}{2a} \log \frac{x - a}{x + a} + C.$

(2) $\int \dfrac{x}{(x^2 + 4x + 5)(x - 1)}\, dx.$

Let $\quad \dfrac{x}{(x^2 + 4x + 5)(x - 1)} = \dfrac{A}{x - 1} + \dfrac{Bx + C}{x^2 + 4x + 5}.$

Proceeding in the ordinary way we find

$$A = \tfrac{1}{10}, \quad B = -\tfrac{1}{10}, \quad C = \tfrac{1}{2}.$$

Therefore

$$I = \frac{1}{10} \int \frac{dx}{x - 1} - \frac{1}{10} \int \frac{x - 5}{x^2 + 4x + 5}\, dx$$

$$= \frac{1}{10} \log (x - 1) - \frac{1}{20} \int \frac{2x + 4}{x^2 + 4x + 5}\, dx + \frac{7}{10} \int \frac{dx}{(x + 2)^2 + 1}$$

$$= \frac{1}{10} \log (x - 1) - \frac{1}{20} \log (x^2 + 4x + 5) + \frac{7}{10} \tan^{-1}(x + 2) + C.$$

(3) $\int \dfrac{x^3}{(x - 1)^2 (2x^2 + x + 1)}\, dx.$

$$\text{Exp.} = \frac{A}{x - 1} + \frac{B}{(x - 1)^2} + \frac{Cx + D}{2x^2 + x + 1}.$$

Clearing of fractions we get

$$x^3 \equiv A (x - 1)(2x^2 + x + 1) + B(2x^2 + x + 1) + (Cx + D)(x - 1)^2.$$

We get B by putting $x = 1$, whence $B = \tfrac{1}{4}$.

Equating the three highest coefficients we get

$$\left. \begin{aligned} 1 &= 2A + C \\ 0 &= -A + 2B - 2C + D \\ 0 &= B + C - 2D \end{aligned} \right\}.$$

Solving these equations

$$A = \tfrac{5}{8}, \quad B = \tfrac{1}{4}, \quad C = -\tfrac{9}{16}, \quad D = -\tfrac{3}{16}.$$

$$\therefore I = \int \frac{5}{8} \cdot \frac{1}{x-1}\, dx + \int \frac{1}{4} \cdot \frac{1}{(x-1)^2}\, dx - \int \frac{2x+3}{16\,(2x^2+x+1)} dx$$

$$= \int \frac{5}{8} \cdot \frac{1}{x-1}\, dx + \int \frac{1}{4} \cdot \frac{1}{(x-1)^2}\, dx - \int \frac{1}{32} \cdot \frac{4x+1}{2x^2+x+1}\, dx$$

$$- \int \frac{5}{32} \cdot \frac{1}{2x^2+x+1}\, dx$$

$$= \frac{5}{8} \log (x-1) - \frac{1}{4} \cdot \frac{1}{x-1} - \frac{1}{32} \log (2x^2+x+1)$$

$$- \int \frac{5}{64} \cdot \frac{dx}{(x+\frac{1}{2})^2+\frac{1}{4}}$$

$$= \frac{5}{8} \log (x-1) - \frac{1}{4} \cdot \frac{1}{x-1} - \frac{1}{32} \log (2x^2+x+1)$$

$$- \frac{5}{128} \tan^{-1}(2x+1) + C.$$

EXAMPLES IV

Integrate the following by the method of the last chapter:

(1) $\dfrac{1}{x^2+4x+3}$.

(2) $\dfrac{x}{x^2+2x+3}$.

(3) $\dfrac{x}{(4-5x)^2}$.

(4) $\dfrac{x}{x^2-5x+6}$.

(5) $\dfrac{2x^2+3x+4}{x^2+6x+10}$.

(6) $\dfrac{(x-1)^2}{x^2+2x+2}$.

*(7) $\dfrac{3}{x^3-3x^2+2}$.

(8) $\dfrac{x^2+1}{x^3+1}$.

(9) $\dfrac{1}{x^3+x^2+x}$.

Find:

(10) $\displaystyle\int_0^1 \frac{x^2+x+1}{x^2-x+1}\, dx$.

(11) $\displaystyle\int \frac{x^2\, dx}{x^4-x^2-12}$.

(12) $\displaystyle\int \frac{dx}{(x^2+a^2)\,(x^2+b^2)}$.

(13) $\displaystyle\int \frac{dx}{x^4+1}$.

(14) $\displaystyle\int \frac{\cos x\, dx}{16-9\sin^2 x}$.

(15) $\displaystyle\int_0^{\frac{\pi}{2}} \frac{\cos x\, dx}{(1+\sin x)\,(5+\sin x)}$.

(16) $\displaystyle\int_0^\infty \frac{x^2\, dx}{(x^2+a^2)\,(x^2+b^2)}$.

CHAPTER V

TRIGONOMETRICAL INTEGRALS
AND FORMULÆ OF REDUCTION

19. Trigonometrical Integrals. Odd power of sine or cosine.

We consider in this chapter various types of trigonometrical integrals frequently met with.

Any odd power of a sine or cosine, or any product of the form $\sin^a x \cos^b x$, where either of the indices a or b is an odd integer, can be evaluated at once by substitution. The method is clear from the following examples.

(a) $\int \sin^7 x\,dx = \int \sin^6 x . \sin x\,dx.$

Let $\qquad z = \cos x, \quad - dz = \sin x\,dx.$

Then $\quad I = - \int (1 - z^2)^3\,dz$

$$= \int (z^6 - 3z^4 + 3z^2 - 1)\,dz$$

$$= \frac{z^7}{7} - \frac{3z^5}{5} + \frac{3z^3}{3} - z + C$$

$$= \tfrac{1}{7} \cos^7 x - \tfrac{3}{5} \cos^5 x + \cos^3 x - \cos x + C.$$

(b) Or again, to integrate $\dfrac{\sin^4 x}{\cos^5 x}$,

$$\int \frac{\sin^4 x}{\cos^5 x} = \int \frac{\sin^4 x}{\cos^6 x} . \cos x\,dx.$$

Let $\qquad z = \sin x, \quad dz = \cos x\,dx.$

Then I becomes $\displaystyle\int \frac{z^4}{(1 - z^2)^3}\,dz.$

Now this is an algebraic fraction and can clearly be evaluated by the methods of a preceding chapter. We need not continue the solution, which is laborious; it is at any rate clear that the integral is no longer trigonometrical.

We see that the general principle is to choose the substitutions so that the function of which there is an odd power is the differential and the integral then becomes algebraic in form. The case when both indices are even presents a greater difficulty and is left till later in the chapter.

20. Integrals of the form

$$\int \frac{dx}{a + b \sin x}, \quad \int \frac{dx}{a + b \cos x}.$$

A second large class of trigonometrical integrals are of the form

$$\int \frac{dx}{a + b \cos x} \quad \text{or} \quad \int \frac{dx}{a + b \sin x}.$$

These, in common with many other trigonometrical integrals, are best dealt with by making the tangent of half the angle the subject of integration.

If $t = \tan \dfrac{x}{2}$, we know that

$$\sin x = \frac{2t}{1 + t^2}, \quad \cos x = \frac{1 - t^2}{1 + t^2}.$$

Also

$$dt = \frac{1}{2} \sec^2 \frac{x}{2} \, dx,$$

i.e.

$$\frac{2dt}{1 + t^2} = dx.$$

The details of this transformation are well worth committing to memory. The use of them will cause the integrals we are considering, and many other trigonometrical integrals, to take an algebraic form, and integration can then be completed by the methods of Ch. IV.

21. Worked Examples.

(1) Find $\int \sec x\,dx$ and $\int \operatorname{cosec} x\,dx$.

Let $t = \tan\dfrac{x}{2}, \quad dx = \dfrac{2dt}{1+t^2}.$

$$\therefore \int \operatorname{cosec} x\,dx = \int \frac{1+t^2}{2t} \cdot \frac{2dt}{1+t^2} = \int \frac{1}{t} \cdot dt$$

$$= \log t = \log \tan\frac{x}{2} + C.$$

Similarly $\int \sec x\,dx = \int \dfrac{1+t^2}{1-t^2} \cdot \dfrac{2dt}{1+t^2}$

$$= \int \frac{2}{1-t^2}\,dt$$

$$= \int \left(\frac{1}{1-t} + \frac{1}{1+t}\right) dt$$

$$= \log \frac{1+\tan\dfrac{x}{2}}{1-\tan\dfrac{x}{2}} + C,$$

or $\qquad\qquad \log \tan\left(\dfrac{\pi}{4} + \dfrac{x}{2}\right) + C.$

We thus have two new standard forms

$$\int \sec x\,dx = \log \tan\left(\frac{\pi}{4} + \frac{x}{2}\right) + C,$$

$$\int \operatorname{cosec} x\,dx = \log \tan\frac{x}{2} + C.$$

(2) $\int \dfrac{dx}{5 + 3\sin x}$.

Making the same substitutions as above we get

$$
\begin{aligned}
I &= \int \dfrac{\dfrac{2dt}{1+t^2}}{5 + \dfrac{6t}{1+t^2}} \\
&= \int \dfrac{2dt}{5 + 6t + 5t^2} \\
&= \dfrac{2}{5} \int \dfrac{dt}{t^2 + \frac{6}{5}t + 1} \\
&= \dfrac{2}{5} \int \dfrac{dt}{(t + \frac{3}{5})^2 + (\frac{4}{5})^2} \\
&= \dfrac{2}{5} \cdot \dfrac{5}{4} \tan^{-1} \dfrac{t + \frac{3}{5}}{\frac{4}{5}} + C \\
&= \tfrac{1}{2} \tan^{-1} \dfrac{5\tan\dfrac{x}{2} + 3}{4} + C.
\end{aligned}
$$

The student will observe that all integrals of this type eventually become of the form $\int \dfrac{1}{z^2 + a^2}\,dz$ or $\int \dfrac{1}{z^2 - a^2}\,dz$ and hence the functions appearing in the answer are usually likely to be inverse tangents or logarithms. The same method and remarks apply to any integral of the form $\int \dfrac{dx}{a + b\cos x + c\sin x}$. But the student will best make himself familiar with the forms that can occur by examples.

EXAMPLES V A

Integrate the following:

(1) $\sin^3 x$.   (2) $\sin^5 x$.   (3) $\cos^9 x$.   (4) $\sin^2 x \cos^5 x$.

(5) $\dfrac{\cos^5 x}{\sin^4 x}$.   (6) $\dfrac{\cos^2 x}{\sin^3 x}$.   (7) $\sin 2x \cos^4 x$.

(8) $\sin 3x \cos^4 x$.   (9) $\dfrac{1}{3+5\cos x}$.   (10) $\dfrac{1}{3-5\sin x}$.

(11) $\dfrac{1}{\sin x + \cos x}$.   (12) $\dfrac{1}{1+\sin x + \cos x}$.   (13) $\dfrac{1}{2+\sin x + \cos x}$.

(14) $\dfrac{\sec x}{a + b \tan x}$.

*(15) Show that

$$\int \frac{\sin x}{\cos 2x}\, dx = \frac{1}{2\sqrt{2}} \log \cot\left(\frac{\pi}{8} + \frac{x}{2}\right) \cot\left(\frac{\pi}{8} - \frac{x}{2}\right) + C.$$

*(16) Prove that

$$\int \frac{\sin\theta\, d\theta}{3\cos\theta + 4\sin\theta} = \frac{4}{25}\theta + \frac{3}{25}\log(3\cos\theta + 4\sin\theta) + C.$$

Find

*(17) $\displaystyle\int_0^{\frac{\pi}{2}} \frac{dx}{1 - \cos a \sin x}$.   (18) $\displaystyle\int_0^{\frac{\pi}{2}} \frac{1}{1 - k^2 \sin^2 x}$.

(19) $\displaystyle\int \frac{dx}{\cos a + \cos x}$.

(20) Prove that, if $a < 1$,

$$\int_0^\pi \frac{dx}{1 - 2a\cos x + a^2} = \frac{\pi}{1 - a^2}.$$

22. Integrals of even order.

In a previous section we postponed the discussion of trigonometrical integrals of even order. There is no general rule for these integrals which can be readily obtained from first principles.

For small values of the indices the integrals may often be evaluated by the aid of multiple angles.

For example $\cos^2\theta = \dfrac{1 + \cos 2\theta}{2}$.

Hence $\displaystyle\int \cos^2\theta\, d\theta = \dfrac{\theta}{2} + \dfrac{\sin 2\theta}{4} + C.$

Or again $\quad \int \sin^2 \theta \cos^2 \theta \, d\theta = \int \frac{1}{4} \sin^2 2\theta \, d\theta$

$$= \int \left[\frac{1}{8} - \frac{\cos 4\theta}{8} \right] d\theta$$

$$= \frac{\theta}{8} - \frac{\sin 4\theta}{32} + C.$$

But although this can always theoretically be done, it is clear that the method is too laborious to be of any practical use.

In certain other cases the substitution $x = \tan \theta$ may be employed.

For example, consider $\int \dfrac{\sin^2 \theta}{\cos^6 \theta} \, d\theta$.

Put $\qquad x = \tan \theta, \quad dx = \sec^2 \theta \, d\theta,$

i.e. $\qquad\qquad \dfrac{dx}{1 + x^2} = d\theta.$

Also $\qquad \sin \theta = \dfrac{x}{\sqrt{1 + x^2}} \quad$ and $\quad \cos \theta = \dfrac{1}{\sqrt{1 + x^2}}.$

Hence $\qquad I = \int \dfrac{x^2}{1 + x^2} \cdot \dfrac{(1 + x^2)^2}{1} \cdot dx$

$$= \int x^2 (1 + x^2) \, dx$$

$$= \int x^2 + x^4 \, dx$$

$$= \frac{x^3}{3} + \frac{x^5}{5} + C.$$

$\therefore \;\; I = \frac{1}{3} \tan^3 \theta + \frac{1}{5} \tan^5 \theta + C.$

Here again it is clear that the scope of the method is limited.

23. Reduction Formulæ.

The discussion of the general case brings us at once to a mention of Reduction Formulæ. Space and the professed scope of this volume will not allow us to deal fully with this subject—but we hope that the student will at least acquire a slight familiarity with the chief forms.

We have already seen how many integrals, themselves apparently insoluble, can be connected in some way or other with other integrals more readily found. In a manner of speaking any such device for connecting one integral with another is a "reduction formula"—for instance, the method of Integration by Parts might be so termed. But the term "reduction formula" is usually restricted to methods of connecting one integral with another of the same type but of lower order, a process which may have to be repeated a number of times before we arrive at an integral easily evaluated. This method is particularly of use in dealing with such forms as $\int \sin^m x\, dx$, $\int \cos^m x\, dx$, $\int \cos^n x \sin^m x\, dx$. The actual reduction formulæ are usually found by trial, as in the following examples.

24. To find reduction formulæ for

$$\int \sin^n x\, dx \quad \text{and} \quad \int \cos^n x\, dx.$$

Consider first $\int \sin^n x\, dx$.

We endeavour to connect this integral with $\int \sin^{n-2} x\, dx$.

Let $P = \sin^{n-1} x \cos x$.

Then $\quad \dfrac{dP}{dx} = (n-1)\sin^{n-2}x\cos^2 x - \sin^{n-1}x \,.\, \sin x$

$$= (n-1)\sin^{n-2}x\,(1-\sin^2 x) - \sin^n x$$

$$= (n-1)\sin^{n-2}x - n\sin^n x.$$

Hence integrating both sides we get

$$P = (n-1)\int \sin^{n-2}x\,dx - n\int \sin^n x\,dx,$$

or $\displaystyle\int \sin^n x\,dx = -\dfrac{1}{n}\{\sin^{n-1}x\cos x\} + \dfrac{n-1}{n}\int \sin^{n-2}x\,dx.$

Similarly, putting $n-2$ for n,

$$\int \sin^{n-2}x\,dx = -\dfrac{1}{n-2}\{\sin^{n-3}x\cos x\} + \dfrac{n-3}{n-2}\int \sin^{n-4}x\,dx,$$

and substituting we can connect

$$\int \sin^n x\,dx \quad\text{with}\quad \int \sin^{n-4}x\,dx.$$

Proceeding in this way we see that $\displaystyle\int \sin^n x\,dx$ can finally be connected *either* with $\displaystyle\int \sin x\,dx$ (if n is an odd integer) *or* with $\displaystyle\int 1\,dx$ (if n is an even integer), and thus the integral can finally be evaluated.

$\displaystyle\int \cos^n x\,dx$ can be found in exactly the same way by choosing $P = \cos^{n-1}x\sin x.$

If n is a negative integer, we can instead connect

$$\int \sin^n x\,dx \quad\text{with}\quad \int \sin^{n+2}x\,dx$$

by choosing $P = \sin^{n+1}x\cos x$ and following the same method; and likewise for $\displaystyle\int \cos^n x\,dx$ when n is negative we choose $P = \cos^{n+1}x\sin x.$

25. Two important definite integrals.

These methods are particularly useful in dealing with certain definite integrals, arising in problems of mensuration, etc. The commonest integrals of this type are those in which the limits are $\frac{\pi}{2}$ and zero or multiples of these limits.

We proceed to show how $\int_0^{\frac{\pi}{2}} \sin^n x\, dx$ and $\int_0^{\frac{\pi}{2}} \cos^n x\, dx$ can be written down at once with the aid of these methods.

We have seen that

$$\int \sin^n x\, dx = -\frac{1}{n}\left\{\sin^{n-1} x \cos x\right\} + \frac{n-1}{n}\int \sin^{n-2} x\, dx,$$

$$\therefore \int_0^{\frac{\pi}{2}} \sin^n x\, dx = \frac{n-1}{n}\int_0^{\frac{\pi}{2}} \sin^{n-2} x\, dx,$$

for the quantity $-\frac{1}{n}\left\{\sin^{n-1} x \cos x\right\}$ vanishes for both limits.

Denote $\int_0^{\frac{\pi}{2}} \sin^n x\, dx$, which is clearly a function of n, by u_n. Then we have successively

$$\left.\begin{array}{l} u_n = \dfrac{n-1}{n}\, u_{n-2} \\[2mm] u_{n-2} = \dfrac{n-3}{n-2}\, u_{n-4} \\[1mm] \text{etc.} \end{array}\right\} \quad \ldots\ldots\ldots\ldots(A).$$

Two cases naturally arise: (1) n odd, (2) n even.

Case I. If n is odd, the last of the equations (A) is

$$u_3 = \tfrac{2}{3} u_1.$$

Hence eliminating all the intermediate functions we get

$$u_n = \frac{n-1}{n} \cdot \frac{n-3}{n-2} \cdots \cdot \frac{4}{5} \cdot \frac{2}{3} u_1.$$

But $\qquad u_1 = \int_0^{\frac{\pi}{2}} \sin x \, dx = \Big[-\cos x \Big]_0^{\frac{\pi}{2}} = 1.$

$\therefore$ Finally $\displaystyle\int_0^{\frac{\pi}{2}} \sin^n x \, dx \, \{n \text{ odd}\} \equiv \frac{n-1}{n} \cdot \frac{n-3}{n-2} \cdots \cdot \frac{2}{3}.$

Case II. If, however, n is even, the last of the relations (A) will be $u_2 = \frac{1}{2} u_0.$

Hence eliminating all intermediate functions we now get

$$u_n = \frac{n-1}{n} \cdot \frac{n-3}{n-2} \cdots \cdot \frac{1}{2} u_0.$$

But $\qquad u_0 = \int_0^{\frac{\pi}{2}} \sin^0 x \, dx = \int_0^{\frac{\pi}{2}} 1 \, dx = \frac{\pi}{2}.$

$\therefore$ Finally

$$\int_0^{\frac{\pi}{2}} \sin^n x \, dx \, \{n \text{ even}\} \equiv \frac{n-1}{n} \cdot \frac{n-3}{n-2} \cdots \cdot \frac{1}{2} \cdot \frac{\pi}{2}.$$

The reader should notice carefully how these results are written down, as it is often useful to be able to quote them.

We place the index n in the first denominator, writing above it $n-1$ as the first numerator. We proceed thus, writing first the denominator and then the numerator till we come to the figure 2. If 2 is a numerator we leave the result as it stands, but if 2 is a denominator we add another factor $\frac{\pi}{2}$. The reader should make himself proficient in this by a few examples of his own choice.

26. $$\int_0^{\frac{\pi}{2}} \cos^n x\, dx = \int_0^{\frac{\pi}{2}} \sin^n x\, dx.$$

It is quite easy to find $\int \cos^n x\, dx$ in the same manner, but it is more instructive to prove that

$$\int_0^{\frac{\pi}{2}} \sin^n x\, dx \quad \text{and} \quad \int_0^{\frac{\pi}{2}} \cos^n x\, dx$$

are equal.

For let $$x = \frac{\pi}{2} - y,$$

i.e. $$dx = -dy,$$

and when the limits for x are $\frac{\pi}{2}$ and 0, those for y are 0 and $\frac{\pi}{2}$.

Hence $$\int_0^{\frac{\pi}{2}} \sin^n x\, dx = \int_{\frac{\pi}{2}}^0 \cos^n y\, (-dy)$$

$$= -\int_{\frac{\pi}{2}}^0 \cos^n y\, dy.$$

But clearly to change the order of the limits changes the sign of the integrals.

$$\therefore \int_0^{\frac{\pi}{2}} \sin^n x\, dx = +\int_0^{\frac{\pi}{2}} \cos^n y\, dy.$$

Hence the results we have just proved for

$$\int_0^{\frac{\pi}{2}} \sin^n x\, dx \quad \text{apply equally to} \quad \int_0^{\frac{\pi}{2}} \cos^n x\, dx.$$

27. Other Limits.

But care must be exercised in applying any of these results when the limits are other than $\frac{\pi}{2}$ and 0; for instance, π and 0.

Thus while it is correct to state that

$$\int_0^\pi \sin^n x \, dx = 2 \int_0^{\frac{\pi}{2}} \sin^n x \, dx,$$

the same result is only true for cosines if n is even; if n is odd the value of $\int_0^\pi \cos^n x \, dx$ being zero.

This is obvious from the fact that the sine repeats its positive values in the second quadrant, whilst the cosine changes sign; or may be proved rigidly as follows:

Now $\quad \int^\pi \cos^n x \, dx = \int_{\frac{\pi}{2}}^\pi \cos^n x \, dx + \int_0^{\frac{\pi}{2}} \cos^n x \, dx.$

Let $\qquad\qquad\qquad y = \pi - x,$

$$dy = - dx.$$

Also if limits for x are π and $\frac{\pi}{2}$, those for y are $0, \frac{\pi}{2}$.

$$\therefore \int_{\frac{\pi}{2}}^\pi \cos^n x \, dx = \int_{\frac{\pi}{2}}^0 \{- \cos y\}^n \cdot - dy$$

$$= \int_0^{\frac{\pi}{2}} (-1)^n \cos^n y \, dy$$

$$= + \int_0^{\frac{\pi}{2}} \cos^n y \, dy \quad \text{if } n \text{ is even,}$$

but $\qquad\qquad\qquad - \int_0^{\frac{\pi}{2}} \cos^n y \, dy \quad \text{if } n \text{ is odd,}$

and hence the whole integral

$$= 0 \quad \text{if } n \text{ is odd,}$$

but $$= 2 \int_0^{\frac{\pi}{2}} \cos^n x \, dx \quad \text{if } n \text{ is even.}$$

The only safe rule for the beginner is to treat every example on its own merits, and, wherever the least doubt arises, make some substitution such as the above to verify his conclusions.

28. The form $\int \sin^p x \cos^q x \, dx$.

The method of successive reductions can also be applied to any integral of the form $\int \sin^p x \cos^q x \, dx$. This can be connected with a number of other integrals of parallel form, the methods being analogous in each case.

We will content ourselves here with exhibiting the connection between

$$\int \sin^p x \cos^q x \, dx \quad \text{and} \quad \int \sin^p x \cos^{q-2} x \, dx.$$

The other reductions will be found in the set of examples at the end of the chapter.

As before we take one more sine and one less cosine, i.e. we start with $P = \sin^{p+1} x \cos^{q-1} x$.

$$\therefore \frac{dP}{dx} = (p+1) \sin^p x \cos^q x - (q-1) \sin^{p+2} x \cos^{q-2} x$$

$$= (p+1) \sin^p x \cos^q x$$
$$\qquad - (q-1) \sin^p x (1 - \cos^2 x) \cos^{q-2} x$$

$$= (p+q) \sin^p x \cos^q x - (q-1) \sin^p x \cos^{q-2} x.$$

$\therefore$ Integrating both sides we have

$$P = (p+q) \int \sin^p x \cos^q x \, dx - (q-1) \int \sin^p x \cos^{q-2} x \, dx.$$

Hence

$$\int \sin^p x \cos^q x\,dx = -\frac{\sin^{p+1} x \cos^{q-1} x}{p+q}$$
$$+ \frac{q-1}{p+q}\int \sin^p x \cos^{q-2} x\,dx.$$

The same process may be applied to the latter integral and we can proceed thus, reducing the order of the integral step by step until either all the cosines have disappeared or only one cosine is left.

In the first case we can go on to evaluate $\int \sin^p x\,dx$ by the methods of §24; in the second case the integral may be solved by the substitution $y = \sin x$.

In practice only the first case is likely to occur, as when either p or q is odd the integral can be dealt with much more easily by substitution, as explained in the earlier part of this chapter.

The connection with $\int \sin^{p-2} x \cos^q x\,dx$ can be effected in an exactly similar manner.

29. The definite integral $\int_0^{\frac{\pi}{2}} \sin^p x \cos^q x\,dx$.

As before, a case which occurs very frequently is that in which p and q are both positive integers, and the limits are $\frac{\pi}{2}$ and 0.

In this case it is easy to see that the reduction formula is simply

$$\int_0^{\frac{\pi}{2}} \sin^p x \cos^q x\,dx = \frac{q-1}{p+q}\int_0^{\frac{\pi}{2}} \sin^p x \cos^{q-2} x\,dx.$$

Again

$$\int_0^{\frac{\pi}{2}} \sin^p x \cos^{q-2} x\, dx = \frac{q-3}{p+q-2} \int_0^{\frac{\pi}{2}} \sin^p x \cos^{q-4} x\, dx,$$

and so on.

Assuming p and q even we get finally

$$\int_0^{\frac{\pi}{2}} \sin^p x \cos^2 x\, dx = \frac{1}{p+2} \int_0^{\frac{\pi}{2}} \sin^p x\, dx$$

$$= \frac{1}{p+2} \cdot \frac{p-1}{p} \cdot \frac{p-3}{p-2} \cdots \frac{1}{2} \cdot \frac{\pi}{2}.$$

Hence finally, eliminating all intermediate integrals, we have

$$\int_0^{\frac{\pi}{2}} \sin^p x \cos^q x\, dx \quad (p \text{ and } q \text{ both even})$$

$$\equiv \frac{(p-1)(p-3)\ldots\ldots 3.1.(q-1)(q-3)\ldots\ldots 3.1}{(p+q)(p+q-2)\ldots\ldots 4.2} \cdot \frac{\pi}{2}.$$

It is evident from the symmetry of the result that

$$\int_0^{\frac{\pi}{2}} \sin^p x \cos^q x\, dx = \int_0^{\frac{\pi}{2}} \sin^q x \cos^p x\, dx,$$

when p and q are both even, a result which can easily be proved directly.

30. In the case where either p or q is negative we connect the integral with one of apparently higher order, e.g. if p is negative we connect

$$\int \sin^p x \cos^q x\, dx \quad \text{with} \quad \int \sin^{p+2} x \cos^q x\, dx,$$

by taking $P = \sin^{p+1} x \cos^{q+1} x$ and differentiating, the steps being exactly as before. By this method the index of $\sin x$ is successively raised by 2 at each reduction until it is finally either zero or unity, in either of which cases integration can be effected by methods mentioned in various parts of this chapter.

In fact, by correctly choosing P, the integral may be connected with any of the integrals

$$\int \sin^{p \pm 2} x \cos^q x \, dx,$$

$$\int \sin^p x \cos^{q \pm 2} x \, dx,$$

$$\int \sin^{p \pm 2} x \cos^{q \pm 2} x \, dx.$$

The reader is left to find the details of these reductions for himself.

31. The definite integrals of § 25 lead to a curious result, first discovered by Wallis.

Since $\quad \sin x < 1, \quad \sin^{n-1} x > \sin^n x > \sin^{n+1} x.$

$$\therefore \int_0^{\frac{\pi}{2}} \sin^{n-1} x \, dx > \int_0^{\frac{\pi}{2}} \sin^n x \, dx > \int_0^{\frac{\pi}{2}} \sin^{n+1} x \, dx.$$

Let n be even and positive.
Then

$$\frac{n-2}{n-1} \cdot \frac{n-4}{n-3} \cdot \ldots \cdot \frac{2}{3} > \frac{n-1}{n} \cdot \frac{n-3}{n-2} \cdot \ldots \cdot \frac{1}{2} \cdot \frac{\pi}{2}$$

$$> \frac{n}{n+1} \cdot \frac{n-2}{n-1} \cdot \ldots \cdot \frac{2}{3}.$$

Hence $\quad \dfrac{\pi}{2} > \dfrac{2^2 \cdot 4^2 \cdot \ldots \cdot (n-2)^2 \cdot n^2}{1 \cdot 3^2 \cdot 5^2 \cdot \ldots \cdot (n-1)^2 (n+1)},$

but $\quad < \dfrac{2^2 \cdot 4^2 \cdot \ldots \cdot (n-2)^2 \cdot n}{1^2 \cdot 3^2 \cdot 5^2 \cdot \ldots \cdot (n-1)^2},$

and since the ratio of these fractions is $\dfrac{n}{n+1}$, which is ultimately unity, the result is sometimes stated in the form

$$\frac{\pi}{2} = \operatorname*{Lt}_{n \to \infty} \frac{2^2 \cdot 4^2 \cdot \ldots (n-2)^2 \cdot n}{1 \cdot 3^2 \cdot 5^2 \cdot \ldots (n-1)^2}.$$

These are known as Wallis' Formulæ.

EXAMPLES V B

Integrate the following:

(1) $\sin^4 x$.　　(2) $\cos^6 x$.　　(3) $\sin^2 x \cos^2 x$.　　(4) $\sin^4 x \cos^2 x$.

(5) $\dfrac{1}{\sin^2 x \cos^2 x}$.　　　　(6) $\dfrac{1}{\sin^3 x}$.　　　　(7) $\dfrac{1}{\sin x \cos^2 x}$.

(8) Find a reduction formula for $\int \sin^n x \, dx$ when n is negative.

(9) Connect $\int \sin^p x \cos^q x \, dx$ with

(a) $\int \sin^{p+2} x \cos^{q-2} x \, dx$,　　　　(b) $\int \sin^{p-2} x \cos^{q+2} x \, dx$.

Write down the values of the following:

(10) $\displaystyle\int_0^{\frac{\pi}{2}} \sin^8 \theta \, d\theta, \quad \int_0^{\pi} \sin^8 \theta \, d\theta, \quad \int_0^{\pi} \cos^8 \theta \, d\theta.$

(11) $\displaystyle\int_0^{\frac{\pi}{2}} \cos^6 \theta \, d\theta, \quad \int_0^{\frac{\pi}{2}} \cos^7 \theta \, d\theta, \quad \int_0^{\pi} \cos^7 \theta \, d\theta, \quad \int_{-\pi}^{\pi} \cos^7 \theta \, d\theta.$

(12) $\displaystyle\int_0^{\frac{\pi}{2}} \sin^4 \theta \cos^4 \theta \, d\theta, \quad \int_0^{\pi} \sin^4 \theta \cos^4 \theta \, d\theta.$

*(13) $\displaystyle\int_0^{\frac{\pi}{2}} (2 \sin^5 x - 3 \sin^7 x) \cos^2 x \, dx.$

(14) Prove the formula

$$\int \sec^{2n+1} \phi \, d\phi = \frac{1}{2n} \tan \phi \sec^{2n-1} \phi + \frac{2n-1}{2n} \int \sec^{2n-1} \phi \, d\phi.$$

4—2

(15) Integrate $\sec^3 \phi$ and $\sec^5 \phi$.

(16) Prove

$$\int x^m (\log x)^p \, dx = \frac{x^{m+1}(\log x)^p}{m+1} - \frac{p}{m+1} \int x^m (\log x)^{p-1} \, dx$$

and integrate $x^4 (\log x)^2$.

(17) Show that $\displaystyle\int_0^a x^2 (a^2 - x^2)^{\frac{3}{2}} = \frac{\pi a^6}{32}$.

(18) Find a reduction formula for

$$\int x^m (a^2 - x^2)^{\frac{p}{2}} \, dx.$$

(19) Show that by taking $P = x^{m+\frac{1}{2}} (2a-x)^{\frac{3}{2}}$ we can find a reduction formula for $\displaystyle\int x^m \sqrt{2ax - x^2} \, dx$.

(20) Calculate $\displaystyle\int_0^{2a} x^4 \sqrt{2ax - x^2} \, dx$.

CHAPTER VI

THE FUNDAMENTAL "INVERSION THEOREM." APPLICATION OF CALCULUS TO FINDING OF AREAS

32. The two meanings of integration.

Up to the present we have defined the process of integration as being the reversal of that of differentiation and it is presumed that, by this time, the student has some idea of the main methods which can be employed in evaluating an integral from this point of view. Looked at in this light, integration is merely an artificial mathematical process, and, did it not possess another meaning, would quickly cease to be of any interest or importance. It is, however, quite reasonable to expect that this mathematical process will have some definitely physical meaning, for the process of which it is the reverse (viz. differentiation) is known to be the method of* finding the "rate of increase" of the given function. We will now proceed to show that the process of integration, or at least that of definite integration, is equivalent to that of finding the sum of a certain series of small quantities.

---

* The student should observe that there is a fundamental theorem of the same character in Differential, viz. that the result of the purely analytical process of finding $\underset{h \to 0}{\text{Lt}} \dfrac{\phi(x+h) - \phi(x)}{h}$ is really the same as that of finding the "rate of increase" of $\phi(x)$. The proof of this theorem being easier than that of the Integral Theorem, we are rather apt to lose sight of its existence.

33. The Fundamental Theorem; Proof by consideration of an Area.

A rough idea of the theorem can be gained from the consideration of the problem of finding the area of the figure bounded by a given curve, two given ordinates, and the axis of x. In Integral Calculus this area is often referred to as the area of (or under) the curve between the given limits.

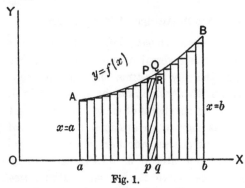

Fig. 1.

Suppose the equation of the bounding curve to be (in ordinary rectangular Cartesians) $y = f(x)$ and the bounding ordinates to be the lines $x = a$, $x = b$.

Divide the base into n equal divisions of width h and complete rectangles as in Fig. 1.

Obviously the sum of the areas of these rectangles constitutes an approximation to the area under the curve and this approximation can be made closer and closer by increasing the number of rectangles (and, of course, decreasing their width h). We may say, in fact, that the area under the curve is the limit of the sum of the areas of an indefinitely large number of rectangles of very small width, formed as in the figure.

Further, we can form an expression for the area of any such rectangle. For if the point of it which rests on the curve be (x_1, y_1) its area is $y_1 h$, i.e. $h \cdot f(x_1)$.

Or, in the limit, we may write this $f(x_1)\,\delta x$, since h is by definition the increase in x in passing from one rectangle to the next.

Hence the whole area is $\Sigma f(x)\,\delta x$, the summation being taken over all possible values of x lying between the limits $x = b$ and $x = a$.

But we may also look at the matter in a slightly different way. Suppose the area already reckoned up to a point $P(x, y)$ to be called $A(x)$, this notation being adopted to remind ourselves that A is a function of x.

Then in passing from P to an adjacent point Q we increase the area by an amount which we naturally denote by $\delta A(x)$. But the actual increase is the portion $PQqp$, which, as we have said, is closely approximated to by the rectangle whose area is $y\,\delta x$.

We have then $\qquad \delta A(x) = y\,\delta x$, approximately,

i.e. $\qquad\qquad \dfrac{\delta A(x)}{\delta x} = y.$

Or, in the limit, when the number of divisions is indefinitely increased,

$$\frac{d A(x)}{dx} = y.$$

Hence according to our definition of integration

$$A(x) = \int y\,dx + C = \int f(x)\,dx + C.$$

Now assume for the moment that $\int f(x)\,dx$ can be evaluated by the previous methods and that the result is $F(x)$.

Then $A(x) = F(x) + C.$

The constant C must clearly be chosen in such a way that when P and A coincide the result is zero, i.e.

$$C = -F(a).$$

Hence the area under the curve up to a point whose abscissa is x is $F(x) - F(a)$.

Hence the area required up to a point whose abscissa is b is $F(b) - F(a)$.

But this would be exactly the value of the definite integral $\int_a f(x)\, dx$, which is thus a second expression for the area under the curve.

These two expressions must, of course, be equal.

34. As an additional verification we can calculate the "apparent error" in this method.

The areas which are apparently neglected in Fig. 1 are a number of nearly triangular portions of which PQR is typical. Each of these "triangles" stands on an equal base and the sum of their respective heights is equal to the distance between the ordinates of A and B.

Hence the sum of all such "triangles" is $\frac{1}{2}h \{Bb \sim Aa\}$.

Now $(Bb \sim Aa)$ is definitely finite and hence the apparent error is a small quantity which $\to 0$ as $h \to 0$.

It is not in fact as large as the "mean" strip, for this would be

$$\tfrac{1}{2}h \{Bb + Aa\}.$$

This argument does not break down even in the case of Fig. 1 a.

For down to the turning point T the errors are of excess, but from T to B of defect. The errors must therefore be reckoned of opposite sign and we are still led to the same result, $\frac{1}{2}h\{Bb\sim Aa\}$, in whatever manner the figure is drawn. The student can verify this by a figure with two turning points between A and B.

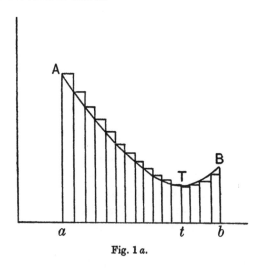

Fig. 1 a.

To give a purely analytical proof of this theorem is a matter of great intricacy and hardly required here. For further information the student is referred to Goursat's *Cours d'Analyse* and Hardy's *Pure Mathematics*. The proof given is, however, valid for any function that is capable of being represented by a continuous graph.

35. The Fundamental Theorem (*continued*).

We have thus arrived at the Fundamental Theorem of the Integral Calculus, which can be applied to integrals of any kind.

The operation $\int_a^b \phi(x)\,dx$ represents the summation of all possible expressions $\phi(a+rh) \cdot h$, where r is any integer, $h = \dfrac{b-a}{n}$ and $n \to \infty$.

Or putting the same statement in the form of a rough rule:

$\int_a^b \phi(x)\,dx$ **is the sum of all expressions like** $\phi(x)\,dx$ between x = b and x = a.

The reader should notice that the method of proof given above implies certain restrictions on the nature of $f(x)$. It assumes, for example, that $f(x)$ is continuous; i.e. that the curve passes from A to B without any abrupt changes in the value of y, and further that no point exists between the end points for which $f(x)$ becomes infinite (though as a matter of fact a value of x for which $f(x) \to \infty$ may be one of the limits itself in certain cases).

This theorem, which seems to have been discovered independently by Newton and Leibnitz, enables us to write down at once the results of summations of this character, many of which are impossible to effect by ordinary algebraical methods of summation. The reader would do well to spend some time and trouble over the proof of this theorem before proceeding further. A thorough understanding of its principles is absolutely necessary to him, if he is to use the Calculus freely.

36. Areas of Curves.

It follows at once from the above work that areas included between any given curve and the axis of x may be calculated from the formula $\int y\,dx$; y being substituted in

terms of x from the equation to the curve and the proper limits taken.

Similarly areas between the given curve and the axis of y may be calculated in the same manner in terms of y from the formula $\int x\,dy$.

The whole area of closed curves may be evaluated by adding or subtracting such areas, as may be seen from the diagrams.

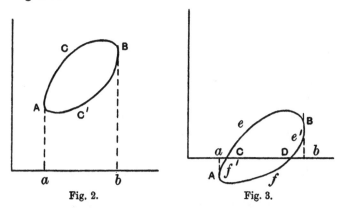

Fig. 2.　　　　　　Fig. 3.

Fig. 2. Area of curve = area $ACBba$ – area $AC'Bba$.

Fig. 3. Area = $CeBbC - De'BbD + DfAaD - Cf'AaC$.

The reader can easily construct other diagrams for himself.

He will also notice that care needs to be exercised in the case of a curve which crosses the axis of x.

If, for example, he were performing the process suggested for Fig. 3, he would find that the portions $DfAaD$, $Cf'AaC$ would both come out with a negative sign (which he would have to disregard in performing the final additions and subtractions). This is easily seen to be the case, since the element $y\,dx$ changes sign at C and D.

In fact, for a portion of a curve crossing the axis between the limits, the ordinary formula gives the difference of the areas formed and not their sum. The only safe rule is to draw the figure and perform the integrations separately.

In the case of symmetrical curves it is often only necessary to find the area of half or a quadrant of the curve. For example, in finding the area of the ellipse $\dfrac{x^2}{a^2} + \dfrac{y^2}{b^2} = 1$, we need find only the area of the positive quadrant (included above the axis of x between $x = a$, $x = 0$) and multiply the result by 4.

37. Parametric Notations.

Many curves, in particular curves of the cycloid family, have their equations most simply expressed in parametric coordinates, i.e. x and y are both given in terms of some third variable t.

The expressions for the area then become $\displaystyle\int y\,\frac{dx}{dt}\,.\,dt$ and $\displaystyle\int x\,\frac{dy}{dt}\,.\,dt$, the proper limits being substituted for t and the whole question worked in terms of t.

Remember that when a curve is given in this form **it is almost always a mistake to attempt to eliminate the parameter. The work is invariably easier with it than without it.**

38. Worked Examples.

(1) Find the area between $y = x^2 + x + 2$, $x = 4$, $x = 1$ and the axis of x.

Noticing first that the curve does not cross the axis we have

$$A = \int_{1}^{4} (x^2 + x + 2)\, dx$$

$$= \left[\frac{x^3}{3} + \frac{x^2}{2} + 2x \right]_{1}^{4}$$

$$= \frac{63}{3} + \frac{15}{2} + 6$$

$$= 34\tfrac{1}{2} \text{ units.}$$

(2) Find the area included between the parabola $= 4ax$, the axis of x and the line $x = h$.

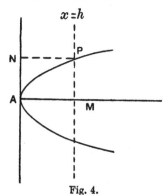

Fig. 4.

It is a little easier to calculate the area APN and to subtract it from the rectangle $AMPN$.

When $x = h$, $y = 2\sqrt{ah}$.

$$\text{Area } APN = \int_{0}^{2\sqrt{ah}} x\, dy$$

$$= \int_{0}^{2\sqrt{ah}} \frac{y^2}{4a}\, dy$$

$$= \frac{1}{4a} \left[\frac{y^3}{3} \right]_{0}^{2\sqrt{ah}}$$

$$= \frac{2}{3} h\sqrt{ah}.$$

But rectangle $AMPN = 2h\sqrt{ah}$.

$\therefore$ Portion $APM = \frac{4}{3}h\sqrt{ah}$.

It is, of course, not difficult to calculate APM directly, but the above method is instructive.

(3) To find the area of the ellipse

$$\frac{x^2}{a^2} + \frac{y^2}{b^2} = 1.$$

The curve consists of four equal quadrants.

The curve is expressible parametrically in the form

$$x = a\cos\theta,$$
$$y = b\sin\theta,$$

and the first quadrant is traced out by the variation of θ from 0 to $\frac{\pi}{2}$.

$\therefore$ Area of 1st quadrant

$$= \int_{\frac{\pi}{2}}^{0} y\,\frac{dx}{d\theta}\,.\,d\theta$$

$$= \int_{\frac{\pi}{2}}^{0} b\sin\theta\,.\,(-a\sin\theta)\,d\theta$$

$$= ab\int_{0}^{\frac{\pi}{2}} \sin^2\theta\,d\theta$$

$$= ab\,.\,\frac{1}{2}\,.\,\frac{\pi}{2} \qquad \{\text{Ch. V}\}$$

$$= \frac{\pi ab}{4}.$$

Hence the whole area of the ellipse is πab.

(4) To find the area between $y^2(a+x) = (a-x)^3$ and its asymptote.

The asymptote clearly is $x = -a$, for as $x \to -a$, $y \to \pm\infty$.

Also the curve cuts the axis where $x = a$, and $x \not> a$, for if $x > a$ or $< -a$, $y^2 < 0$.

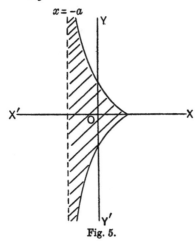

Fig. 5.

Hence the curve is as in Fig. 5.

Area is given by $2 \int_{-a}^{a} \dfrac{(a-x)^{\frac{3}{2}}}{(a+x)^{\frac{1}{2}}} dx.$

To evaluate put $x = a \sin \theta.$

The limits are now $\pm \dfrac{\pi}{2}$ and $dx = a \cos \theta \, d\theta.$

$$\therefore A = 2 \int_{-\frac{\pi}{2}}^{\frac{\pi}{2}} a^2 \frac{(1 - \sin \theta)^{\frac{3}{2}}}{(1 + \sin \theta)^{\frac{1}{2}}} . \cos \theta \, d\theta$$

$$= 2 \int_{-\frac{\pi}{2}}^{\frac{\pi}{2}} a^2 \frac{(1 - \sin \theta)^{\frac{3}{2}} (1 - \sin^2 \theta)^{\frac{1}{2}}}{(1 + \sin \theta)^{\frac{1}{2}}} \, d\theta$$

$$= 2a^2 \int_{-\frac{\pi}{2}}^{\frac{\pi}{2}} (1 - \sin \theta)^2 \, d\theta$$

$$= 2a^2 \left[\theta + 2\cos\theta + \frac{\theta}{2} - \frac{\sin 2\theta}{4} \right]_{-\frac{\pi}{2}}^{\frac{\pi}{2}}$$

{expanding and integrating}

$$= 3\pi a^2 \text{ finally.}$$

39. Simpson's Approximate Rule for Area.

Consider the parabola $y = ax^2 + bx + c$.

The area included between this curve, the axis of x and two ordinates $x = h_1, h_2$, is clearly equal to

$$\frac{a}{3}\{h_2^3 - h_1^3\} + \frac{b}{2}\{h_2^2 - h_1^2\} + c\{h_2 - h_1\},$$

which may be written in the form

$$\{h_2 - h_1\}\left\{\frac{a}{3}\{h_2^2 + h_2 h_1 + h_1^2\} + \frac{b}{2}\{h_2 + h_1\} + c\right\}.$$

Now the ordinate at $x = h_1$ is clearly of length

$$ah_1^2 + bh_1 + c,$$

that at $x = h_2$ $\qquad ah_2^2 + bh_2 + c,$

and that at the mid-point $x = \dfrac{h_1 + h_2}{2}$

$$\frac{a}{4}\{h_1^2 + 2h_1 h_2 + h_2^2\} + \frac{b}{2}\{h_1 + h_2\} + c,$$

and we thus see that the area given above can be expressed as

$$\frac{h_2 - h_1}{6}\{1\text{st ordinate} + 4 \text{ times middle ordinate}$$

$$+ 2\text{nd ordinate}\}\ldots(A).$$

By considering every three consecutive points of a curve as if joined up by some parabola of the form given above (which we clearly have a right to do) we get Simpson's Rule for the Approximate Area under a curve. This is a result which is often useful in obtaining the area of a graph whose equation is not given.

The rule is as follows:

Divide the area by any <u>odd</u> number of equidistant ordinates. The area is then closely represented by the expression {one-third of interval between ordinates} × {1st ordinate + last ordinate + twice sum of all other odd ordinates + 4 times sum of all even ordinates}.

The student will easily prove this expression by repeating equation (A) the required number of times and adding.

EXAMPLES VI

(1) Find the area included between $y = 3x^2 + x + 3$, the axis of x and the ordinates $x = 2$, $x = 1$.

Also the area between the same curve, the axis of y and corresponding lines drawn perpendicular to it.

(2) Obtain the areas under the following curves between $x = 0$ and $x = h$:

      (a) $y^3 = a^2 x$;

      (b) $a^2 y = x^3 + ax^2$.

(3) Calculate the area of one arch of the curve $y = \sin x$.

(4) Find the total area included between the two parabolas $y^2 = 4ax$ and $x^2 = 4ay$.

(5) Use the Calculus to find the area of the circle $x^2 + y^2 = a^2$.

(6) Explain why the result of Ex. 5 cannot be regarded as a valid proof of the formula for the area of a circle.

(7) Find the area of the loop of the curve $y^2 = x^2 + x^3$.

(8) A cycloid is given by the equations
$$x = a\{\theta + \sin\theta\},$$
$$y = a\{1 - \cos\theta\}.$$

Make a rough figure of the curve, and calculate the area of one complete arch.

(9) Prove that the whole area of the curve $\left\{y - \dfrac{x^2}{a}\right\}^2 = a^2 - x^2$ is πa^2.

*(10) Find the ratio in which the curve $ay^2 = x^3$ divides the rectangle formed by drawing perpendiculars from any point on the two axes.

(11) Find the area above the axis of x bounded by the curve
$$y = (x^2 - 1)(4 - x^2).$$

(12) For the curve $y^2(a - x) = x^2(a + x)$ find

    (i)   the area of the loop;

    (ii) the area between the curve and its asymptote.

(13) Find the whole area between $x^2y^2 = a^2(y^2 - x^2)$ and its asymptotes.

CHAPTER VII

FURTHER APPLICATIONS TO GEOMETRY: SECTORIAL AREAS, VOLUMES OF REVOLUTION, FINDING THE LENGTH OF A CURVE

40. Areas in Polar Coordinates.

It is sometimes necessary to calculate areas belonging to curves whose equations are expressed in Polar Coordinates. The areas usually required are "sectorial areas," i.e. areas included between the curve and two given radii vectores.

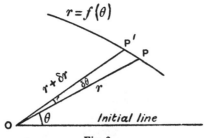

Fig. 6.

The formula for such areas can easily be found as follows: Consider two close points on the curve. In the limit when P and P' are on the point of coinciding we may regard the arc PP' as indistinguishable from its chord (cf. Fig. 5, which is a portion of a circle of large radius); if, then, P is (r, θ) and P' $(r + \delta r, \theta + \delta \theta)$,

the area $OPP' = $ area of $\triangle OPP' = \frac{1}{2} OP \cdot OP' \sin POP'$
$$= \frac{1}{2} r (r + \delta r) \sin \delta \theta.$$

But since a small angle and its sine are ultimately equal

this reduces to $\frac{1}{2} r (r + \delta r) \delta \theta$, and neglecting the term $\frac{1}{2} r \delta r \delta \theta$, which is of the second order, we have simply

$$\frac{1}{2} r^2 \delta \theta.$$

Hence, since the area to be determined consists of a large number of such elements, it can by the fundamental theorem be found by evaluating $\frac{1}{2} \int_{a}^{\beta} r^2 d\theta$, where $\theta = \beta$ and $\theta = \alpha$ are the two radii vectores which bound the area.

41. Example.

Consider the whole area of the curve $r = a(1 + \cos\theta)$ [the Cardioid]. The curve consists of two equal portions

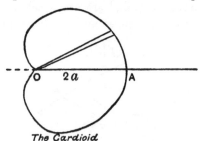

The Cardioid

Fig. 7.

as shown. The upper portion is traced by the revolution of the radius vector from 0 to π.

$$\therefore \text{Area} = 2 \int_0^\pi \tfrac{1}{2} r^2 d\theta$$

$$= \int_0^\pi a^2 (1 + \cos\theta)^2 d\theta$$

$$= a^2 \int_0^\pi (1 + 2\cos\theta + \cos^2\theta) d\theta$$

$$= a^2 \left[\theta + 2\sin\theta + \frac{\theta}{2} + \frac{\sin 2\theta}{4} \right]_0^\pi$$

$$= \frac{3\pi a^2}{2}.$$

42. Volumes of Revolution.

The Calculus can also be employed to calculate the volumes of solids bounded by known surfaces. In general the process requires a knowledge of triple integration, but the simpler case of volumes of revolution can be dealt with by a single integration.

Consider the solid formed by the revolution about the axis of x of the portion of the curve $y = f(x)$ considered in § 33.

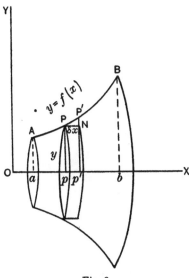

Fig. 8.

Dividing the area into strips as before we see that each strip, in revolving about OX, contributes to the volume a thin plate. In the limit we can regard $PP'p'p$ as a rectangle and the volume contributed by the revolution of this rectangle will be $\pi \cdot Pp^2 \cdot pp'$, or using the usual notation of the Calculus $\pi y^2 \delta x$.

Applying the fundamental theorem as before, we can find the sum of all these circular plates by evaluating $\int_a^b \pi y^2 \, dx$.

In the same way a volume of revolution about OY would be given by the formula $\int \pi x^2 \, dy$.

Also if x and y are given parametrically we can write the formula for a volume of revolution $\int \pi y^2 \cdot \dfrac{dx}{dt} \, dt$ and evaluate in terms of the parameter t.

43. Worked Examples.

(1) To find the volume of a sphere.

Consider the revolution of one quadrant of the circle

$$x^2 + y^2 = a^2.$$

In the first quadrant x ranges from a to 0.

$$\therefore \text{Volume contributed} = \pi \int_0^a (a^2 - x^2) \, dx$$

$$= \pi \left\{ a^2 x - \frac{x^3}{3} \right\}_0^a$$

$$= \frac{2\pi a^3}{3}.$$

$$\therefore \text{Whole volume} = \tfrac{4}{3}\pi a^3.$$

(2) To find the volume of a cone of height h and base radius r.

The cone can be considered as formed by the revolution of the line $y = \dfrac{r}{h} x$ about Ox.

$$\therefore \text{Volume} = \int_0^h \pi \frac{r^2}{h^2} x^2 dx$$

$$= \pi \frac{r^2}{h^2} \cdot \left(\frac{x^3}{3}\right)_0^h$$

$$= \frac{\pi r^2 h}{3}.$$

(3) To find the volume of revolution formed by the loop of the curve $y^2 = x^2 \dfrac{a-x}{a+x}$.

[The whole curve lies between $\pm a$, for otherwise y^2 is negative, also at the origin $y^2 = x^2$ and $y \to \pm \infty$ as $x \to -a$. Hence curve is as in figure.]

Volume traced by loop

$$= \pi \int_0^a y^2 dx$$

$$= \pi \int_0^a x^2 \cdot \frac{a-x}{a+x} dx.$$

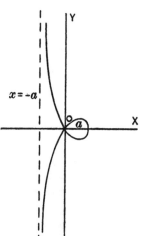

Fig. 9.

Put $z = x + a$, then $dz = dx$ and new limits are $2a$ and a.

$\therefore$ Volume

$$= \pi \int_a^{2a} \frac{(z-a)^2 (2a-z)}{z} dz$$

$$= \pi \int_a^{2a} \left[\frac{2a^3}{z} - 5a^2 + 4az - z^2\right] dz$$

$$= \pi \left[2a^3 \log z - 5a^2 z + 2az^2 - \frac{z^3}{3}\right]_a^{2a} = 2\pi a^3 \{\log 2 - \tfrac{2}{3}\}.$$

EXAMPLES VII A

(1) Find the area of a loop of the curve $r = a \sin 4\theta$.

(2) Find the whole area of $r^2 = a^2 \cos^2 \theta + b^2 \sin^2 \theta$.

(3) Find the area of one loop of $r^2 = a^2 \cos 2\theta$. (The Lemniscate.)

(4) Find the area of the Limaçon $r = a + b \cos \theta$ when $a > b$.

(5) Modify the result of Ex. 4 to cover the cases (i) $b = a$, (ii) $a < b$.

(6) Find the area enclosed between the spiral $r = ae^\theta$, the initial line and the radius $\theta = \pi$.

(7) Find the volume of the "frustum" of a cone, if the height of the frustum is h and the radii of its ends r_1 and r_2.

(8) Find the volume cut off from a sphere of radius a by two parallel planes which meet the sphere in circles of radii r_1 and r_2.

(9) Find the volume of the "spheroid" traced out by the revolution of the ellipse $\dfrac{x^2}{a^2} + \dfrac{y^2}{b^2} = 1$.

(10) Assuming the earth's section to be an ellipse of eccentricity $\frac{1}{60}$, the minor axis being the line from the centre to the North Pole, calculate the percentage error in taking for its volume the volume of the sphere whose radius is that from the centre to the equator.

*(11) Find the volume formed by the revolution about Ox of the portion of the parabola $y^2 = 4ax$ bounded by $x = h$, and prove that this volume is $\frac{2}{3}$ of that of the enveloping cone formed by the tangents at the points where $x = h$.

(12) Find the volume formed by the revolution about Ox of the portion of the cycloid $x = a(\theta + \sin \theta)$, $y = a(1 - \cos \theta)$ given by values of θ ranging from 0 to π.

*(13) Find the volume of the segment of a sphere of radius r cut off by a plane at a distance $r - c$ from the centre.

*(14) A cylindrical hole is drilled symmetrically through a hemisphere at right angles to its plane surface. The radius of the hole being $\frac{2}{3}$ that of the sphere, find the volume of the portion removed.

*(15) A solid is generated by the revolution of $y^2 = Ax^2 + Bx + C$ about the axis of x. Prove that the volume cut off by two planes perpendicular to Ox at a distance h apart is $\frac{1}{6}h\{A_1 + 4A + A_2\}$, where A_1, A_2 are the end-areas and A the area of the section half-way along.

*(16) The segment of the parabola $y^2 = 4ax$ cut off by $x = h$ revolves about that ordinate. Prove that the volume so formed is $\frac{8}{15}$ of that of the cylinder formed by the revolution of the tangent at the vertex.

*(17) Find the position of a point on the curve $ay^2 = x^3$ such that, if perpendiculars are drawn from it to the axes, the volumes formed by the revolution of the two parts about their appropriate axes are equal.

44. Finding the Length of an Arc.

The Calculus may also be employed to find the length of an arc of any curve whose equation is given.

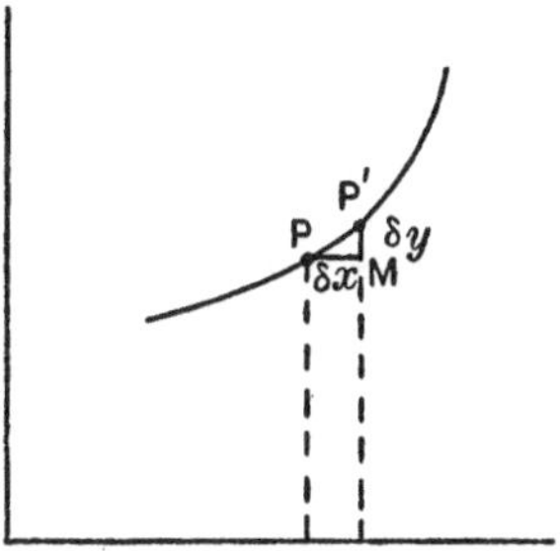

For if P and P' are two close points of the curve we may in the limit regard the chord PP' as indistinguishable from the arc PP'.

But chord

$$PP' = \sqrt{P'M^2 + PM^2},$$

i.e. $\delta s^2 = \delta x^2 + \delta y^2.$

Fig. 10.

Various formulæ for s can at once be derived from this.

For considering x as independent variable we have, dividing by δx^2 and taking the limit,

$$\frac{ds}{dx} = \sqrt{1 + \left(\frac{dy}{dx}\right)^2},$$

$$\therefore s = \int \sqrt{1 + \left(\frac{dy}{dx}\right)^2}\, dx,$$

the integration being taken between limits corresponding to the ends of the arc to be measured.

Or, again, if y is considered as the primary variable we get

$$\frac{ds}{dy} = \sqrt{1 + \left(\frac{dx}{dy}\right)^2},$$

or

$$s = \int \sqrt{1 + \left(\frac{dx}{dy}\right)^2}\, dy.$$

Most useful of all is the "parametric" form

$$\left(\frac{ds}{dt}\right)^2 = \left(\frac{dx}{dt}\right)^2 + \left(\frac{dy}{dt}\right)^2,$$

whence

$$s = \int \sqrt{\left(\frac{dx}{dt}\right)^2 + \left(\frac{dy}{dt}\right)^2}\, dt.$$

It is assumed in all cases that the curve is continuous between the two ends.

45. Length. Formulæ in Polar Coordinates.

Or again, in Polar coordinates, if OP and OP' be two

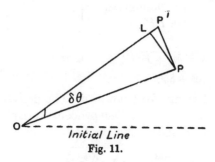

Fig. 11.

close radii vectores and PL perpendicular to OP', we have the arc PP' approximately equal to the chord PP'.

But $$PP'^2 = PL^2 + LP'^2.$$

In the limit when $\delta\theta$ is a small angle we may consider
$$OL = OP,$$
$$\therefore LP' = r + \delta r - r = \delta r,$$
and also LP approx. = arc of a circle of radius OP
$$= r\,\delta\theta.$$
$$\therefore \delta s^2 = \delta r^2 + r^2 \delta\theta^2.$$

This in the same manner as in the preceding article gives us the results

$$s = \int \sqrt{1 + r^2\left(\frac{d\theta}{dr}\right)^2}\, dr,$$

and
$$s = \int \sqrt{r^2 + \left(\frac{dr}{d\theta}\right)^2}\, d\theta.$$

The process of finding s is often called "rectification," or "rectifying the arc."

46. Worked Examples.

(1) To find the length of the parabola $y^2 = 4ax$ included between the vertex and the point where $x = a$.

Choosing y as the primary variable the limits are $y = 2a$, $y = 0$.

$\therefore$ By formula

$$s = \int_0^{2a} \sqrt{1 + \left(\frac{dx}{dy}\right)^2}\, dy$$

$$= \int_0^{2a} \sqrt{1 + \left(\frac{y}{2a}\right)^2}\, dy \qquad \left\{\text{for } \frac{dy}{dx} = \frac{2a}{y}\right\}$$

$$= \frac{1}{2a} \int_0^{2a} \sqrt{y^2 + 4a^2}\, dy$$

$$= \frac{1}{2a} \left[\frac{y\sqrt{y^2 + 4a^2}}{2} + \frac{4a^2}{2} \log \frac{y + \sqrt{y^2 + 4a^2}}{2a} \right]_0^{2a}$$

$$= \frac{1}{4a} \left[2a\sqrt{8a^2} + \frac{4a^2}{2} \log(1 + \sqrt{2}) \right]$$

$$= a\{\sqrt{2} + \log(\sqrt{2} + 1)\}.$$

(2) Find the whole length of the cardioid

$$r = a\,(1 + \cos\theta).$$

We have noticed that the curve is symmetrical, the upper half being traced by the variation of θ from 0 to π.

Also
$$\frac{dr}{d\theta} = -a\sin\theta.$$

$$
\begin{aligned}
\therefore \ \text{Length} &= 2\int_0^\pi \sqrt{r^2 + \left(\frac{dr}{d\theta}\right)^2}\, d\theta \\
&= 2\int_0^\pi \sqrt{a^2(1+\cos\theta)^2 + a^2\sin^2\theta}\, d\theta \\
&= 2a\int_0^\pi \sqrt{2 + 2\cos\theta}\, d\theta \\
&= 2a\int_0^\pi 2\cos\frac{\theta}{2}\, d\theta \\
&= \left[8a\sin\frac{\theta}{2}\right]_0^\pi \\
&= 8a.
\end{aligned}
$$

The integrals given by these formulæ are often of great complexity: the rectification of so simple a curve as the ellipse is, indeed, not expressible without the invention of new functions. Only a few examples are given herewith; and for further information the student is referred to the standard text-books on the Calculus.

EXAMPLES VII B

(1) In the "catenary" $2y = c\left\{e^{\frac{x}{c}} + e^{-\frac{x}{c}}\right\}$, prove that the arc from $x = a$ to $x = 0$ is $\dfrac{c}{2}\left\{e^{\frac{a}{c}} - e^{-\frac{a}{c}}\right\}$.

(2) Find the length of any arc of the spiral $r = a\theta$.

(3) Find the length of a complete arch (from $\theta = 2\pi$ to $\theta = 0$) of the cycloid $x = a\,(\theta + \sin\theta)$, $y = a\,(1 - \cos\theta)$.

(4) Find the length of arc of the curve given oy

$$x = (a+b) \cos \theta - b \cos \frac{a+b}{b} \theta,$$

$$y = (a+b) \sin \theta - b \sin \frac{a+b}{b} \theta,$$

from the point $\theta = \dfrac{\pi b}{a}$ to the point $\theta = \theta_1$.

(5) Find the length from the origin to the point where $x = a$ in the curve $y = a \log \dfrac{a^2}{a^2 - x^2}$.

(6) Obtain a formula for the length of any arc of the curve $ay^2 = x^3$.

7) Find the whole length of the curve $x^{\frac{2}{3}} + y^{\frac{2}{3}} = a^{\frac{2}{3}}$.

CHAPTER VIII

APPLICATION TO PROBLEM OF FINDING CENTRES OF GRAVITY

47. Centre of Gravity. Principle.

We often require in Mechanics to determine the position of the centre of gravity of a given body, plane or solid. The coordinates of the C.G. can always be expressed by means of the Calculus, if the equation to the bounding curve or surface is known.

We shall confine our treatment to two cases, viz. Areas, and Surfaces of Revolution, since these are the only cases which do not in general involve double or triple integration.

The principle invoked is the Statical Principle of Moments, viz. that if a body be divided in any manner, the algebraical sum of the moments of the separate portions about any axis is equal to the moment about that axis of the whole body supposed collected at its C.G.

48. To determine the Centre of Gravity of an Area.

Consider first the problem of finding the C.G. of an area. Regard the area as a thin plane lamina of uniform surface density σ; and suppose that the area is that bounded by the curve $y = f(x)$, the axis of x and the two ordinates $x = a$, $x = b$. We have previously seen that any area can be obtained by adding or subtracting two or more such areas.

To determine the x coordinate of the C.G. we proceed as follows: Divide the area into strips parallel to the axis of y. The area of a typical strip will be $y \, \delta x$ and its mass $\sigma y \, \delta x$.

Hence the moment of such a strip about Oy will be $\sigma xy\,\delta x$, for every part of the strip is (to the first order) at a distance x from Oy.

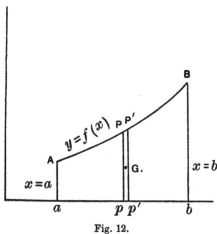

Fig. 12.

Applying the fundamental theorem the sum of all such moments will be $\int_a^b \sigma xy\,dx$.

But the whole area is $\int_a^b y\,dx$ and hence the whole mass $= \sigma \int_a^b y\,dx$.

Hence if $(\bar{x}, \bar{y})$ be the C.G. we have

$$\bar{x}\,.\,\sigma \int_a^b y\,dx = \sigma \int_a^b xy\,dx,$$

i.e.

$$\bar{x} = \frac{\int_a^b xy\,dx}{\int_a^b y\,dx}.$$

Again, since any strip is approximately rectangular, its c.g. is at a height $\frac{y}{2}$ above Ox.

$\therefore$ Moment of any strip about $Ox = \sigma \frac{y^2}{2} \delta x$.

$\therefore$ Sum of all such moments $= \sigma \int_a^b \frac{y^2}{2} dx$.

To determine $\bar{y}$ we have the equation

$$\bar{y} \cdot \sigma \int_a^b y \, dx = \sigma \int_a^b \frac{y^2}{2} dx.$$

$$\therefore \bar{y} = \tfrac{1}{2} \frac{\displaystyle\int_a^b y^2 \, dx}{\displaystyle\int_a^b y \, dx}.$$

The student will find it wise, at any rate in the earlier stages of his knowledge, to remember the method rather than the formula and to treat each example on its own merits *ab initio*.

49. We may, in just the same manner, find the c.g. of a volume of revolution. It is clear however in this case that the c.g. lies on the axis of revolution, and hence, in general, there is only one coordinate to be determined. To fix our ideas let us suppose that Ox is the axis of revolution, the solid is that considered in § 42 and the figure is Fig. 8.

The solid can, as in that article, be considered as formed of a large number of circular slabs each generated by the revolution of one of the elementary rectangles. Let ρ be the density.

The volume of any such slab $= \pi y^2 \delta x$.

$\therefore$ Its moment about $Oy = \pi \rho \, y^2 x \delta x$.

$\therefore$ Sum of all such moments $= \pi \rho \int_a^b y^2 x \, dx$.

But the whole mass $= \pi\rho \int_a^b y^2 dx$ and we thus get the equation

$$\bar{x} \cdot \pi\rho \int_a^b y^2 dx = \pi\rho \int_a^b xy^2 dx.$$

$$\therefore \bar{x} = \frac{\int_a^b xy^2 dx}{\int_a^b y^2 dx}.$$

It is seen that the method can easily be extended to cover cases of variable density, as long as the density involves x only. And again it is the method that is worth remembering, not the formula.

50. This is by far the easiest way of finding the C.G. of the simpler bodies as the following examples will show.

(i) Find the C.G. of the area between $y^2 = 4ax$, $y = 0$. and $x = a$.

Dividing into strips parallel to Oy, the total moment of the strips is seen to be

$$\sigma \int_0^a xy \, dx$$

$$= \sigma \int_0^a x \cdot 2\sqrt{ax} \, dx$$

$$= 2\sigma \sqrt{a} \int_0^a x^{\frac{3}{2}} \, dx$$

$$= \tfrac{4}{5}\sigma a^3.$$

But the whole mass $= \sigma \int_0^a y \, dx$

$$= 2\sigma \sqrt{a} \int_0^a x^{\frac{1}{2}} dx$$

$$= \tfrac{4}{3}\sigma a^2.$$

$$\therefore \tfrac{4}{3}\sigma a^2 \cdot \bar{x} = \tfrac{4}{5}\sigma a^3,$$

$$\therefore \bar{x} = \tfrac{3}{5}a.$$

Again, since the ordinate of the c.g. of each strip is $\frac{y}{2}$ the moment of all the strips about Ox is

$$\sigma \int_0^a \frac{y^2}{2}\, dx$$
$$= \tfrac{1}{2}\sigma \int_0^a 4ax\, dx$$
$$= \sigma a^3.$$
$$\therefore \tfrac{4}{3}\sigma a^2 . \bar{y} = \sigma a^3,$$
$$\therefore \bar{y} = \frac{3a}{4}.$$

The c.g. is thus $\left\{\dfrac{3a}{5},\ \dfrac{3a}{4}\right\}$.

(ii) Find the c.g. of a solid cone.

The cone can, as we have seen, be traced out by the revolution of the line $y = \dfrac{r}{h} x$.

Dividing the cone into circular slabs perpendicular to Ox, the moment of any slab about $Oy = \pi\rho x y^2 \delta x$.

$\therefore$ Total moment of all such slabs

$$= \int_0^h \pi\rho x y^2\, dx$$
$$= \pi\rho \frac{r^2}{h^2} \int_0^h x^3\, dx$$
$$= \pi\rho \frac{r^2}{h^2} \cdot \frac{h^4}{4}$$
$$= \tfrac{1}{4}\pi\rho r^2 h^2.$$

But we have proved that the volume $= \tfrac{1}{3}\pi r^2 h$.

$$\therefore \tfrac{1}{3}\pi\rho r^2 h . \bar{x} = \tfrac{1}{4}\pi\rho r^2 h^2,$$
$$\therefore \bar{x} = \frac{3h}{4},$$

i.e. the c.g. is one quarter of the way up the axis of the cone.

(iii) To find the C.G. of a segment of a paraboloid of revolution formed by the revolution of $y^2 = 4ax$ about Ox, the density at any point varying as x^2, i.e. $\rho = kx^2$.

Proceeding as before

the mass of a slab is $\pi k x^2 y^2 \delta x$,

and the moment of a slab is $\pi k x^3 y^2 \delta x$.

Hence $\qquad \bar{x} . \int_0^h \pi k x^2 y^2 \, dx = \int_0^h \pi k x^3 y^2 \, dx,$

i.e. $\qquad \bar{x} . \int_0^h 4ax^3 \, dx = \int_0^h 4ax^4 \, dx,$

i.e. $\qquad\qquad \bar{x} . h^4 = \frac{4h^5}{5} .$

$$\therefore \bar{x} = \frac{4h}{5} .$$

(iv) To find the surface and C.G. of a hollow hemisphere.

Consider the revolution of an element of arc of a quadrant of the circle $x^2 + y^2 = a^2$.

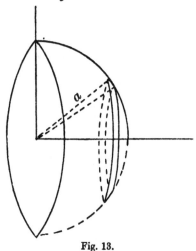

Fig. 13.

The element of arc traces out a belt, and the sum of all such belts is the surface of the hemisphere.

Clearly the radius of the circle traced out by the belt at a point (x, y) is y.

And hence the area of such a belt is $2\pi y . \delta s$. Hence the whole area is $\int 2\pi y \, ds$ taken over limits which include the whole quadrant.

For the circle $x = a \cos \theta$, $y = a \sin \theta$.

Hence $\qquad \dfrac{ds}{d\theta} = \sqrt{\left(\dfrac{dx}{d\theta}\right)^2 + \left(\dfrac{dy}{d\theta}\right)^2} = a,$

i.e. $\qquad ds = a \, d\theta,$

and the whole quadrant is covered if θ varies from 0 to $\dfrac{\pi}{2}$.

$\therefore$ Integral becomes $\displaystyle\int_0^{\frac{\pi}{2}} 2\pi a \sin \theta . a \, d\theta$

$$= 2\pi a^2 \int_0^{\frac{\pi}{2}} \sin \theta \, d\theta$$

$$= 2\pi a^2.$$

Again, to find the C.G.; clearly the C.G. of any belt is at its centre.

$\therefore$ Moment of all belts

$$= \sigma \int 2\pi yx \, ds$$

$$= 2\pi\sigma \int_0^{\frac{\pi}{2}} a^3 \sin \theta \cos \theta \, d\theta$$

$$= \pi\sigma a^3 \int_0^{\frac{\pi}{2}} \sin 2\theta \, d\theta$$

$$= \pi\sigma a^3 \left[-\tfrac{1}{2} \cos 2\theta \right]_0^{\frac{\pi}{2}}$$

$$= \pi\sigma a^3.$$

∴ We have $2\pi\sigma a^2 . \bar{x} = \pi\sigma a^3$.

Hence $\bar{x} = \dfrac{a}{2}$ and clearly $\bar{y} = 0$.

51. Pappus' Theorems.

The following are two theorems often attributed to Guldinus but also found in the works of the Greek geometer Pappus.

(i) *If any closed area revolves about an axis external to the area, the volume formed is equal to the area multiplied by the length of the path traced out by the c.g. of the area.*

(ii) *The surface area so formed is equal to the perimeter of the enclosing curve multiplied by the path of the c.g.*

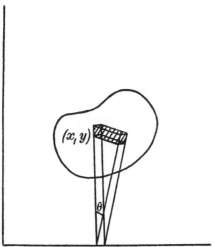

Fig. 14.

A proof of the first of these theorems will suffice.

Consider an element of the area, δA, and suppose that it rotates through an angle $\delta\theta$ about Ox.

The arc traced out by any point within the element will be to the first order $y\,\delta\theta$ and hence the volume contributed by the element is $y\,\delta A\,\delta\theta$.

Or in fact for a rotation through a finite angle α the element will be $y\alpha\,\delta A$, and thus the whole volume will be given by $\int y\alpha\,dA$, the integration being taken over such limits as will cover the whole area.

But, by our previous work, the ordinate of the c.g. can be found from the equation

$$\bar{y} \times \text{Total Mass} = \text{Total Moment},$$

i.e.
$$\bar{y} \times \int dA = \int y\,dA.$$

$$\therefore \bar{y}\,.\,A = \int y\,dA.$$

$\therefore$ Volume $= \int y\alpha\,dA = \alpha\int y\,dA = \bar{y}\alpha\,.\,A$, and $\bar{y}\alpha$ is the path traced by the c.g.

Therefore the theorem is as stated.

In particular for a whole revolution $V = 2\pi\bar{y}\,.\,A$.

It is seen that a similar proof gives the second theorem.

52. Volumes of Revolution. Polar Coordinates.

Pappus' Theorem can be employed to give the volume of revolution formed by an area given in polar coordinates. For we have previously seen that an element of area is $\frac{1}{2}r^2\,\delta\theta$, and, regarding this as a triangle, its c.g. is $\frac{2}{3}$ of the way along the median, i.e. at a point whose coordinates are approximately $\left\{\dfrac{2r\cos\theta}{3},\ \dfrac{2r\sin\theta}{3}\right\}$.

$\therefore$ Path of c.g. $= \frac{4}{3}\pi r\sin\theta$.

$\therefore$ Volume contributed by an element $= \frac{2}{3} \pi r^3 \sin \theta \, \delta\theta$.

$\therefore$ Whole volume is given by $\int \frac{2}{3} \pi r^3 \sin \theta \, d\theta$ between suitable limits.

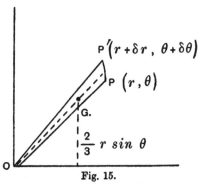

Fig. 15.

Ex. Find the volume traced out by revolution of
$$r = a + b \cos \theta \quad (a > b)$$
about the initial line.

As above
$$V = \frac{2}{3} \pi \int_0^\pi r^3 \sin \theta \, d\theta$$
$$= \frac{2}{3} \pi \int_0^\pi (a + b \cos \theta)^3 \sin \theta \, d\theta.$$

Let
$$u = a + b \cos \theta,$$
$$du = - b \sin \theta \, d\theta.$$

And limits for u are $a - b$, $a + b$.

$$\therefore V = \frac{2\pi}{3} \int_{a+b}^{a-b} - u^3 \cdot \frac{du}{b}$$
$$= \frac{2\pi}{3b} \left\{ \frac{(a+b)^4}{4} - \frac{(a-b)^4}{4} \right\}$$
$$= \frac{2\pi}{12b} \cdot [2a^2 + 2b^2] [4ab]$$
$$= \frac{4\pi}{3} a (a^2 + b^2).$$

If $a < b$, the limits for θ are 0 and $\cos^{-1}\left(-\frac{a}{b}\right)$.

EXAMPLES VIII

Find the c.g. of:

(1) a semi-circle;

(2) the circumference of a semi-circle;

(3) a solid hemisphere;

(4) a quadrant of the ellipse $\dfrac{x^2}{a^2} + \dfrac{y^2}{b^2} = 1$.

(5) Show that the area bounded by the curve $r = f(\theta)$ and the radii vectores $\theta = a$, $\theta = \beta$ has its c.g. at the point

$$x = \frac{2}{3} \frac{\displaystyle\int_a^\beta r^3 \cos\theta \, d\theta}{\displaystyle\int_a^\beta r^2 \, d\theta}, \quad y = \frac{2}{3} \frac{\displaystyle\int_a^\beta r^3 \sin\theta \, d\theta}{\displaystyle\int_a^\beta r^2 \, d\theta}.$$

(6) Prove that the c.g. of the area enclosed by the cardioid $r = a(1 + \cos\theta)$ is at $\left(\dfrac{5a}{6}, 0\right)$.

(7) Find the c.g. of the area between $y^2 = 4ax$ and $y = mx$.

(8) Find the position of the c.g. of the portion of a sphere intercepted between two parallel planes.

Find also an expression for the surface-area of this zone.

(9) Find the c.g. of a frustum of a solid cone, h being the height of the frustum and r, R the radii of its ends.

(10) Find the surface and c.g. of a hollow cone.

(11) Find the c.g. of a straight rod of length l in which the density of an element is proportional to the distance of the element from one end of the rod.

(12) The part of the curve $y^{m+n} = a^m x^n$ between $x = 0$ and $x = h$ revolves about Ox. Find the c.g. of the solid so formed.

(13) Give expressions for the volume and total surface of the "anchor-ring" formed by the revolution of a circle of radius a about an axis in its plane distant d from its centre.

(14) A square revolves about a line through one corner parallel to a diagonal. Find the volume and total surface so formed.

(15) Find the volume traced out by the revolution about the initial line of the loop of the curve $r^3 = a^3 \theta \cos \theta$.

(16) The area between the cardioid $r = 2a(1 + \cos \theta)$ and the parabola $r(1 + \cos \theta) = 2a$ revolves about the initial line. Prove that the volume formed is $18\pi a^3$.

(17) A cotton reel is formed by the revolution of the given figure

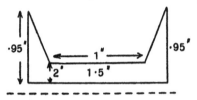

Fig. 16.

about an axis $\frac{1}{10}''$ from its straight edge. Find the volume and total surface of the reel.

*(18) In Ex. 15, VII A, prove that the c.g. is at a distance

$$\frac{h}{2} \frac{A_2 - A_1}{A_2 + 4A + A_1}$$

from one end.

(19) By considering the revolution of a semi-circular area about a diameter find the position of the c.g. of a semi-circle by Pappus' Theorem.

(20) By considering the surface formed by the revolution of one quarter of the circumference of a circle, deduce the c.g. of one quarter of the circumference of a circle.

*(21) An arc of a circle of radius c subtending an angle $2a$ at the centre is rotated about its chord. Prove that the area of the surface of revolution so formed is $4\pi c^2 (\sin a - a \cos a)$.

(22) Deduce from Ex. 21 the position of the c.g. of any circular arc.

CHAPTER IX

FURTHER APPLICATIONS TO MECHANICS. RIGID DYNAMICS

53. Moments of Inertia.

If a particle of mass m lies at a distance r from a fixed axis, the product mr^2 is known as the "moment of inertia" of the particle about that axis. This quantity, as we shall see, figures largely in questions of rotation of a heavy body, and the calculation of moments of inertia is thus one of the most important practical applications of the Calculus. In general we are dealing with solid bodies and not with particles, and we thus have to consider our rigid bodies as composed of a large number of strips, or other convenient divisions, and obtain the sum of the resulting series of quantities by the Calculus. Only in some of the simpler cases may this be effected by a single integration.

There are two general theorems relating to moments of inertia which are often of use in the evaluation and which we proceed to give.

54. Two Important Results on Moments of Inertia.

A. If I_x and I_y are the moments of inertia about two perpendicular axes OX and OY, of a plane lamina lying in the plane XOY, and I_z the moment of inertia of the same lamina about OZ which is perpendicular to the plane XOY, then

$$I_z = I_x + I_y.$$

For consider an element of the body of mass m, lying at a point (x, y) on the plane.

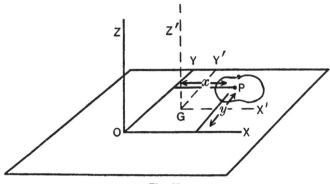

Fig. 17.

Then moment of inertia of the element about OX
$$= mx^2,$$
and that about OY $\qquad = my^2,$
i.e. $I_x = \Sigma mx^2$ and $I_y = \Sigma my^2$, summing the elements.
$$\therefore \ I_x + I_y = \Sigma m\,(x^2 + y^2) = \Sigma m\,.\,OP^2$$
$$= I_z,$$
$$\therefore \ I_z = I_x + I_y.$$

B. If I_A is the moment of inertia of any body of mass M about any axis and I_G the M.I. about a parallel axis through the C.G., these two axes being a distance h apart, then
$$I_A = I_G + Mh^2.$$

First suppose the body is a lamina as in the previous theorem and OZ is the axis under consideration, and let the C.G. be at $(\bar{x}, \bar{y})$.

If now parallel axes GX', GY', GZ' are taken through

G and (x, y) are the old, (x', y') the new coordinates of P, we have
$$x = x' + \bar{x},$$
and
$$y = y' + \bar{y}.$$

Now the moment of inertia about OZ
$$= \Sigma m (x^2 + y^2)$$
$$= \Sigma m (x'^2 + 2x'\bar{x} + \bar{x}^2 + y'^2 + 2y'\bar{y} + \bar{y}^2)$$
$$= \Sigma m (x'^2 + y'^2) + \Sigma m (\bar{x}^2 + \bar{y}^2) + 2\Sigma mx'\bar{x} + 2\Sigma my'\bar{y}$$
$$= \Sigma m (x'^2 + y'^2) + (\bar{x}^2 + \bar{y}^2) \Sigma m + 2\bar{x} \Sigma mx' + 2\bar{y} \Sigma my',$$

bringing outside certain constant terms.

Now $\Sigma m (x'^2 + y'^2)$ is clearly the moment of inertia about the axis GZ'; also $\bar{x}^2 + \bar{y}^2 = OG^2 = h^2$.

The two terms $\Sigma mx'$, $\Sigma my'$ must both vanish for they clearly represent the moments of the lamina about the two axes GX', GY' which contain the c.g.

Hence we get
$$\mathbf{I_{OZ} = I_{GZ} + Mh^2.}$$

Considering a solid body as made up of a number of parallel laminæ we see that the theorem is also true for a solid body.

55. Reason for existence of "Moments of Inertia."

It is of interest at this stage to consider how the Moment of Inertia comes to play the important part it does in Mechanics. Consider a rigid body of any sort rotating about a fixed axis. Any particle of the body at a distance r from the centre will be describing a circle of radius r.

The tangential velocity of this particle will thus be $r\dfrac{d\theta}{dt}$ (θ is the angle described) and hence the tangential acceleration will be the derivative of this, viz. $r\dfrac{d^2\theta}{dt^2}$.

The "effective" force along the tangent will thus be $mr \dfrac{d^2\theta}{dt^2}$ and the moment of this force $mr^2 \dfrac{d^2\theta}{dt^2}$.

Now by d'Alembert's Principle the sum of all such moments must be equal to the moment of the external forces which are giving rise to the motion. Let this moment be denoted by L and we now have the equation

$$\Sigma mr^2 \frac{d^2\theta}{dt^2} = L.$$

But for any rigid body $\dfrac{d^2\theta}{dt^2}$ is constant for all particles composing the body and may be taken outside the summation sign, and Σmr^2 is the quantity we call I (the moment of inertia); we thus arrive at the fundamental equation for the rotation of a fixed body

$$\mathbf{I \frac{d^2\theta}{dt^2} = L.}$$

56. Analogies between Linear and Rotational Dynamics.

The student will notice the close analogy between this equation and that of ordinary Dynamics, $P = mf$.

In fact we may say in general that, in Rotational Dynamics, I takes the place of m in the same way that angular velocity and acceleration take the place of linear velocity and acceleration. For instance it is easily verified that $\frac{1}{2} I \left(\dfrac{d\theta}{dt} \right)^2$ is the energy possessed by a body in virtue of its rotation, and $I \dfrac{d\theta}{dt}$ its "angular" momentum.

Corresponding to the two equations of ordinary Dynamics:

(a) Change of K.E. = Force × Space described,

(b) Change of Momentum = Force × Time,

we have two similar equations in Rigid Dynamics:

(a) Change in value of $\frac{1}{2} I \left(\frac{d\theta}{dt} \right)^2$ = Moment of couple producing change × Angle (in radians) turned through,

(b) Change in value of $I \frac{d\theta}{dt}$ = Moment of couple × Time.

By means of these equations most of the elementary questions in Rigid Dynamics can be solved. An example is given later.

57. Worked Examples.

A few worked examples are given here. For fuller information the student is referred to standard works on Rigid Dynamics.

(1) Find the M.I. of a uniform rod of length $2a$ about its centre.

Let λ be the mass of the rod per unit length. Consider an element of the rod at a distance x from O.

The M.I. of this element about its centre is $\lambda x^2 \delta x$.

Hence by the principles of the Calculus we have to find

$$\int_{-a}^{a} \lambda x^2 \, dx,$$

$$\therefore \ I = \left[\frac{\lambda x^3}{3} \right]_{-a}^{a} = \frac{2\lambda a^3}{3}.$$

But $2\lambda a = M$ (the mass of the rod),

$$\therefore \ I = \frac{Ma^2}{3}.$$

The quantity k which is such that the moment of inertia of the body $= Mk^2$ is often called the "radius of gyration"; e.g. in the last example the radius of gyration of the rod about its centre is $\dfrac{a}{\sqrt{3}}$ *.

(2) We can see similarly (using Theorem B) that the M.I. about one end is $\frac{4}{3}Ma^2$, and that the M.I. about an axis perpendicular to the rod and distant b from its mid-point is

$$M\left(\frac{a^2}{3} + b^2\right).$$

(3) The M.I. of a circular area about any diameter.

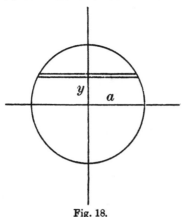

Fig. 18.

Taking the diameter as axis of x the length of a strip at a distance y from Ox is $2\sqrt{a^2 - y^2}$, and since every portion of this strip is at a distance y from Ox, the M.I. of the strip will evidently be $2y^2\sqrt{a^2 - y^2}\,.\,\lambda\,.\,dy.$ (λ = density.)

$$\therefore \ \text{Whole M.I.} = 2\int_{-a}^{a} y^2\sqrt{a^2 - y^2}\,dy.$$

* Physically it is the distance at which a particle of equal mass must be placed to rotate in the same manner as the given body.

Put
$$y = a \sin \theta,$$
$$dy = a \cos \theta \, d\theta,$$

and the new limits are $\pm \dfrac{\pi}{2}$.

$$\therefore \text{ M.I.} = 2 \int_{-\frac{\pi}{2}}^{\frac{\pi}{2}} a^4 \sin^2 \theta \cos^2 \theta \, d\theta$$

$$= \frac{\lambda a^4}{2} \int_{-\frac{\pi}{2}}^{\frac{\pi}{2}} \sin^2 2\theta \, d\theta$$

$$= \frac{\lambda a^4}{4} \int_{-\frac{\pi}{2}}^{\frac{\pi}{2}} (1 - \cos 4\theta) \, d\theta$$

$$= \frac{\lambda a^4}{4} \left[\theta - \frac{\sin 4\theta}{4} \right]_{-\frac{\pi}{2}}^{\frac{\pi}{2}}$$

$$= \frac{\pi \lambda a^4}{4}.$$

But
$$M = \pi \lambda a^2.$$
$$\therefore \text{ M.I.} = \frac{Ma^2}{4}.$$

It follows by symmetry that the Moment of Inertia about Oy also equals $\dfrac{Ma^2}{4}$. And since, by Theorem A, $I_z = I_x + I_y$, we find the Moment of Inertia of a circle about an axis through its centre perpendicular to its plane to be $\dfrac{Ma^2}{2}$.

(4) The student should carefully notice the following worked example of a Dynamical character:

A uniform wheel of mass 100 lbs. and radius of gyration about its centre 1 foot is acted upon by a couple whose moment $= 10$ ft. lb. units for one minute. Find the angular velocity produced. Find also the constant couple which would stop the wheel in half a minute when it is revolving at 15 revs. per second and how many revolutions would be made before coming to rest.

$$\text{M.I.} = Mk^2 = 100 \text{ and } L = 10g \text{ units.}$$

$\therefore$ the equation is $100 \dfrac{d^2\theta}{dt^2} = 10g.$

Integrating both sides $100 \dfrac{d\theta}{dt} = 10gt + C$, an equation giving the angular velocity at time t.

Clearly this velocity must be zero when $t = 0$.

$$\therefore C = 0 \text{ and } \frac{d\theta}{dt} = \frac{gt}{10}.$$

$\therefore$ when $t = 60$ secs. $\dfrac{d\theta}{dt} = 6g$, i.e. an angular velocity of $6g$ radians per sec. is produced.

Again, in the second part of the question, let $-L$ be the required retarding couple.

$$\therefore 100 \frac{d^2\theta}{dt^2} = -L,$$

and integrating $\qquad 100 \dfrac{d\theta}{dt} = -Lt + C.$

But at zero time

$$\frac{d\theta}{dt} = 30\pi \quad (15 \text{ revs. per sec.}),$$

$$\therefore C = 3000\pi.$$

$$\therefore 100 \frac{d\theta}{dt} = -Lt + 3000\pi.$$

If $\dfrac{d\theta}{dt}$ becomes zero at end of 30 secs. we have

$$30L = 3000\pi,$$

i.e. $\qquad\qquad L = 100\pi \text{ units,}$

i.e. required retarding couple is one whose moment is $\dfrac{100\pi}{g}$ ft. lbs.

For the last part of the question we have

$$100\,\frac{d^2\theta}{dt^2} = -100\pi,$$

i.e.

$$\frac{d^2\theta}{dt^2} = -\pi.$$

Multiply both sides by $\dfrac{d\theta}{dt}$ and integrate with respect to t,

i.e.

$$\frac{d\theta}{dt}\cdot\frac{d^2\theta}{dt^2} = -\pi\,\frac{d\theta}{dt},$$

and

$$\frac{1}{2}\left(\frac{d\theta}{dt}\right)^2 = -\pi\theta + C.$$

But when $\dfrac{d\theta}{dt} = 30\pi$, $\theta = 0$, for we are measuring the angle from the initial position given.

$$\therefore\ \frac{900\pi^2}{2} = C.$$

$$\therefore\ \frac{1}{2}\left(\frac{d\theta}{dt}\right)^2 = -\pi\theta + \frac{900\pi^2}{2}.$$

$\therefore$ when the wheel comes to rest, $\dfrac{d\theta}{dt} = 0$ and $\theta = 450\pi$,

i.e. the wheel has revolved 225 times.

The student should observe particularly the last method of integrating the given equation, which is often of use.

EXAMPLES IX A

Answers should be expressed in the form Mk^2.

(1) Find the Moment of Inertia of a rectangle whose sides are $2a$, $2b$, about axes through its centre parallel to its sides.

(2) Hence deduce the Moment of Inertia of a rectangular solid whose sides are $2a$, $2b$, $2c$, about principal axes through its centre.

Find the Moment of Inertia of each of the following:

(3) The circumference of a circle about a diameter.

(4) A thin hollow sphere about any diameter.

(5) An ellipse about its major and minor axes.

(6) A right circular cylinder about its axis.

(7) A circular disc about any tangent.

(8) A thin circular ring about any tangent.

(9) An isosceles triangle about the principal altitude.

(10) A hollow spherical shell of external radius b and internal radius a about a diameter.

(11) A flywheel of mass 500 lbs. has a radius of gyration of $2\frac{1}{2}$ feet. What is its energy when rotating at 200 revs. per minute? If its motion is opposed by a couple whose moment is 12 ft. lbs., find how long will elapse before the wheel comes to rest and how many revolutions it will have made.

(12) Find in the previous question the time in which the velocity will be reduced to half its initial value.

(13) A uniform trap-door of mass M and dimensions 2 feet by 3 feet turns about its smaller edge from an initially horizontal position. Write down its kinetic and potential energies when it has fallen through an angle θ, and deduce the velocity with which it reaches its lowest position.

58. Work done by a varying force.

The Calculus can be employed to find the work done by a varying force, when the law of variation of the force is known, and can be expressed in terms of some parameter x; for let δs be the displacement in the direction of the line of action of the force, corresponding to a small increment δx in the value of the parameter. The work done during this displacement is $P\delta s$ and hence the whole work can be calculated by finding $\int P ds$, the limits being such as to include the whole interval which is being considered.

As an example let us consider the problem of finding the work done in stretching an elastic string from its natural length a to a new length b.

When the length of the string is x ($x > a$) by Hooke's Law the tension will be $\lambda \dfrac{x-a}{a}$, where λ is some constant depending on the composition of the string.

In stretching to a length $x + \delta x$, the tension may be considered as remaining constant and the work done is $\lambda \dfrac{x-a}{a} \delta x$.

$$\therefore \text{ Total work done} = \int_a^b \lambda \frac{x-a}{a}\, dx$$
$$= \frac{\lambda}{a}\left[\frac{(x-a)^2}{2}\right]_a^b$$
$$= \frac{\lambda}{2a}(b-a)^2.$$

But the final tension at length $b = \lambda \dfrac{b-a}{a}$.

Hence the work done $= \frac{1}{2}$ Extension $\times$ Final Tension.

59. Practical Application.

In practical work this calculation of work done is often performed graphically. We can see that if we have a graph in which Force is plotted against the distance the point of application has moved, the area under this curve will represent the work done between any two positions of the machine, for work done $= \displaystyle\int_{x=a}^{x=b} P\,dx =$ area of the (P, x) curve between $x = b$, $x = a$.

For example, in the question considered in the preceding article $T = \lambda \dfrac{x-a}{a}$.

The graph would thus be a straight line inclined at an angle $\tan^{-1}\lambda$ and passing through $(a, 0)$. The work done is thus represented by the area of the $\triangle ATB$ and is clearly $\frac{1}{2}TB . AB$, or $\frac{1}{2}$ Final tension × Extension. Devices are often introduced by which the (P, x) curve can be drawn

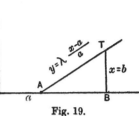

Fig. 19.

automatically by the machine performing the work and the area of these diagrams gives the work done.

Such devices are generally used with machines which perform a " cycle "; doing useful work in one part of the cycle, and requiring work done on them during the other part to restore the machine to its initial position. In a steam engine for instance work is done by the piston during part of the revolution but against the piston in restoring it to its original position.

The work diagram (known as an Indicator diagram) would be roughly of the shape shown in the sketch. The area $PAQNM$ represents the positive work done in the stroke, the area $PA'QNM$ the (negative) work taken up in restoring the piston to the

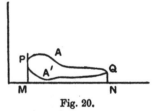

Fig. 20.

zero position. Hence the closed area $PAQA'$ represents the total effective work done per stroke.

60. Total Resultant Thrust and Centre of Pressure.

The Calculus is also used in hydrostatical problems. The pressure of a liquid is a variable force—proportional to the distance below the free surface of the body, or part of a body, on which the pressure acts. It is often desirable to calculate the Total Force (Total Resultant Thrust) on an unequally immersed body—and also to find the point (Centre of Pressure) at which this Total Thrust may be assumed to act.

Simple cases are obviously able to be evaluated by one integration, though more complicated cases often require double integration. The principle is again the Principle of Moments, moments being taken about the free surface of the liquid. A simple example is given herewith.

Ex. A square of side a is immersed vertically, with one edge in the free surface. Find the Total Resultant Thrust and the Centre of Pressure.

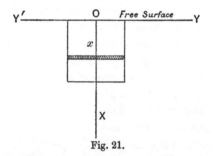

Fig. 21.

Taking the axis of y in the free surface, consider a strip parallel to the free surface at a depth x below it.

The area of the strip is $a\delta x$ and all along it the pressure will be ρgx units of force per unit of area.

$$\therefore \text{ Total Thrust on strip} = \rho g a x \delta x.$$

$$\therefore \text{ Total Thrust on square} = \int_0^a \rho g a x\, dx$$
$$= \frac{\rho g a^3}{2}.$$

Now the moment of the thrust on the strip about Oy

$$= \rho g\, ax^2\, \delta x.$$

$$\therefore \text{Total moment} = \int_0^a \rho g a x^2\, dx$$

$$= \frac{\rho g a^4}{3}.$$

$\therefore$ If Resultant thrust acts at a depth $\bar{x}$ we have

Moment of total thrust $=$ Total moment of thrusts on strips,

i.e.
$$\frac{\rho g a^3}{2}\,\bar{x} = \frac{\rho g a^4}{3},$$

$$\therefore \bar{x} = \frac{2a}{3}.$$

The centre of pressure is at $\frac{2}{3}$ of the depth.

EXAMPLES IX B

(1) A curve is drawn showing the velocity of a particle against the time. Explain how to read from the graph the distance described in a given interval.

Apply your result to the formula $v = u + ft$ and deduce graphically $s = ut + \frac{1}{2}ft^2$.

(2) A volume v of a gas at pressure p is enclosed in a cylindrical vessel of radius r. It is allowed to expand so that its volume is trebled. Find the total work done.

(3) A cork 3 ins. long is drawn from a bottle by a force proportional to the area remaining in contact with the bottle. If the initial pull $= 48$ lbs. weight, find the work done in extracting the cork.

(4) Extend the formula for work done in stretching an elastic string to cover the case in which the string is stretched from a length a (not its natural length) to a length b.

(5) A string 1 foot long is such that if one pound weight be suspended from it an extension of one inch is produced. Find the work done in stretching the string to 15 inches length.

(6) In the previous question the 1 lb. weight is allowed to fall momentarily through 1 foot before the string is taut; to what length will the string be stretched momentarily?

(7) A semi-circle is immersed vertically with its diameter in the free surface. Find the centre of pressure.

(8) A cubical box of side a has a hinged lid. The box is filled with water and tilted about the edge opposite to the hinge. Explain how to find the maximum angle through which we can tilt the cube without the water spilling.

CHAPTER X

DIFFERENTIAL EQUATIONS. A FEW TYPES

61. Differential Equations. General principles.

Equations containing $\frac{dy}{dx}$, $\frac{d^2y}{dx^2}$, ... and functions of y and x are known as Differential Equations. Such equations are of very frequent occurrence in mechanical and physical problems and a large branch of Higher Mathematics concerns itself with them. The determination of the nature of the relation between y and x which allows a given differential equation to be true is known as "solving the equation" or "finding the integral" of the equation. Space prohibits a long discussion of these equations, for further details of which the reader is referred to the standard text-books*.

One general principle becomes evident. It is clear that if the equation contains $\frac{d^n y}{dx^n}$, the solution must be equivalent to the performance, actual or concealed, of n successive integrations. Since each of these integrations introduces an arbitrary constant, the general solution of a differential equation of the nth order will contain n arbitrary constants.

Or the matter may be seen thus. Suppose the equation $f(x, y) = 0$ contains n undetermined constants. We can then differentiate n times, getting the n further equations

$$f_1(x, y) = 0, \quad f_2(x, y) = 0 \text{ and so on,}$$

---

* And in particular to Prof. H. T. Piaggio's excellent *Elementary Treatise on Differential Equations*.

each containing some or all of the arbitrary constants. We now have in all $(n+1)$ equations containing these n constants and therefore we can in theory, at any rate, eliminate them. The eliminant will be an equation containing functions of x, y and derivatives of y up to and including the nth. That is, it will be a differential equation of the nth order; and so, conversely, we may expect that the solution of such an equation will lead us back to a relation containing n different arbitrary constants.

62. Types of solution already considered.

The process of integration is itself, of course, the same as finding the solution of an equation $\dfrac{dy}{dx} = f(x)$, and consequently any equation in which $\dfrac{dy}{dx}$ may be found in terms of x only will only need to be integrated directly, e.g. the equation

$$\left(\frac{dy}{dx}\right)^2 - x\,\frac{dy}{dx} + 6x^2 = 0$$

gives at once
$$\frac{dy}{dx} = 3x \text{ or } -2x,$$

$$\therefore\ y = \text{either } \tfrac{3}{2}x^2 + A \text{ or } -x^2 + B.$$

Similarly the equation $\dfrac{dy}{dx} = f(y)$ can be solved at once by inverting both sides, e.g.

$$\frac{dy}{dx} = \frac{y^2}{1+y},$$

$$\therefore\ \frac{dx}{dy} = \frac{1+y}{y^2} = \frac{1}{y^2} + \frac{1}{y},$$

$$\therefore\ x = -\frac{1}{y} + \log y + \text{const.}$$

And in general any equation in which the variables can be separated can be integrated as in the following example:

$$\frac{dy}{dx} = \frac{\sqrt{1-y^2}}{1+x^2},$$

$$\therefore \frac{dy}{\sqrt{1-y^2}} = \frac{dx}{1+x^2}.$$

Integrating both sides

$$\sin^{-1} y + C = \tan^{-1} x.$$

63. Linear Equations with Constant Coefficients.

We pass on to the consideration of a type which frequently occurs in Mechanical and Physical Problems.

The typical equation is

$$p_0 \frac{d^n y}{dx^n} + p_1 \frac{d^{n-1} y}{dx^{n-1}} + \ldots + p_n y = 0.$$

All the coefficients $p_0 \ldots p_n$ are constants, and no derivative appears except in the first power.

The simplest possible equation of this type is the equation

$$\frac{dy}{dx} = ky,$$

i.e.

$$\frac{dx}{dy} = \frac{1}{ky},$$

i.e.

$$x = \frac{1}{k} \log y + C.$$

For convenience let the arbitrary constant be written $\dfrac{-\log A}{k}$,

$$\therefore x = \frac{1}{k} \log \frac{y}{A},$$

$$\therefore \mathbf{y = Ae^{kx}}.$$

Notice that, as we expect, the solution contains one arbitrary constant.

64. Graphical representation of this result.

The reader should notice the two types of curve which can be represented by this equation according as k is positive or negative. If k is positive we have a curve in which y is always increasing at a rate proportional to its value at the instant considered. This type of relation is sometimes called "True Compound Interest." The graph is as in Fig. 22, the upper curve for a positive and the lower for a negative value of A, and

$$y \to \pm \infty \quad \text{as} \quad x \to \infty,$$
$$y \to 0 \quad \text{as} \quad x \to -\infty.$$

If k is negative the conditions are reversed, giving the two curves of Fig. 23. These are sometimes called "die-away curves."

The reader will do well to consider for himself the various types represented by $y = Ae^{kx} + B$.

65. Equations of the Second Order. The Auxiliary Equation.

We next consider equations of the form

$$p \frac{d^2y}{dx^2} + q \frac{dy}{dx} + ry = 0.$$

We shall expect that this type of equation will occur in Mechanics, for forces and accelerations are expressible in terms of second differentials.

The results found in the previous section suggest that exponential functions will form the basis of the solution.

Let us try as a solution $y = Ae^{ax}$.

Substituting we get

$$Ae^{ax} \{p\alpha^2 + q\alpha + r\} = 0.$$

Hence Ae^{ax} will be a solution when α is either of the roots of the "auxiliary" equation

$$pa^2 + q\alpha + r = 0.$$

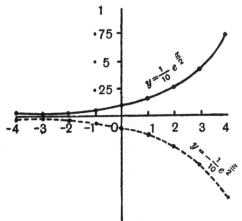

Form $y = Ae^{kx}$. k positive.

Fig. 22.

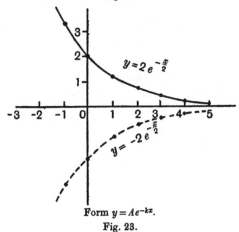

Form $y = Ae^{-kx}$.

Fig. 23.

Denoting these roots by α and β, $Ae^{\alpha x}$ and $Be^{\beta x}$ are both solutions and hence $Ae^{\alpha x} + Be^{\beta x}$ is seen to be a solution containing two arbitrary constants, i.e. a general solution.

The principle can clearly be extended to equations of any order.

$Ex.$ $$\frac{d^2 y}{dx^2} + 7\frac{dy}{dx} + 10y = 0.$$

Auxiliary equation

$$\alpha^2 + 7\alpha + 10 = 0, \quad \text{roots} -5, -2.$$

$\therefore$ Solution is $\quad y = Ae^{-5x} + Be^{-2x}.$

66. Exceptional cases. Imaginary Roots.

It very often happens that the roots of the auxiliary equation are imaginary (whenever $q^2 < 4r$).

The general solution now reduces to the form

$$y = Ae^{(\alpha + \sqrt{-1}\beta)x} + Be^{(\alpha - \sqrt{-1}\beta)x}$$

or $\qquad = e^{\alpha x}\{Ae^{\sqrt{-1}\beta x} + Be^{-\sqrt{-1}\beta x}\}.$

But by Euler's Exponential Values,

$$\left. \begin{array}{l} 2\cos\beta x = e^{\sqrt{-1}\beta x} + e^{-\sqrt{-1}\beta x} \\ 2\sqrt{-1}\sin\beta x = e^{\sqrt{-1}\beta x} + e^{-\sqrt{-1}\beta x} \end{array} \right\} *.$$

Thus any expression $Ae^{\sqrt{-1}\beta x} + Be^{-\sqrt{-1}\beta x}$ can also be written in the form $A_1\cos\beta x + B_1\sin\beta x$, where A_1 and B_1 are two new arbitrary constants. The general solution in the case of imaginary roots is therefore expressed in the form

$$\mathbf{y = e^{\alpha x}\{A\cos\beta x + B\sin\beta x\}.}$$

The case most frequently met with is that where the middle term of the equation is missing, leaving only the terms

$$\frac{d^2 y}{dx^2} + ky = 0 \quad (k \text{ positive}).$$

* Hobson's *Trigonometry*, §§ 228, 229.

The roots being $\pm \sqrt{-k}$, the solution is

$$y = A \cos \sqrt{k}x + B \sin \sqrt{k}x,$$

a form which should be noted for future reference.

67. Equal Roots.

It may also occur that the auxiliary equation has equal roots; and the form of the general solution reduces to $(A + B) e^{ax}$, which only includes one true arbitrary constant and thus is not the general solution.

Consider the equation

$$\frac{d^2y}{dx^2} - 2a \frac{dy}{dx} + a^2 y = 0.$$

The auxiliary equation has equal roots a, and the solution clearly involves e^{ax}.

Put
$$y = e^{ax} . v.$$

Substituting, the equation becomes

$$e^{ax} . \frac{d^2v}{dx^2} = 0,$$

$$\therefore \frac{d^2v}{dx^2} = 0,$$

and
$$v = Ax + B.$$

Hence the general solution is

$$y = e^{ax} (Ax + B).$$

It is clear that this method can be extended to equations of any order.

Ex. (i).
$$\frac{d^3y}{dx^3} + y = 0,$$

roots of $a^3 + 1$ are $- 1, -\frac{1}{2} \pm \frac{\sqrt{-3}}{2}$.

The solution is

$$y = A e^{-x} + e^{-\frac{x}{2}} \left(B \cos \frac{\sqrt{3}\,x}{2} + C \sin \frac{\sqrt{3}\,x}{2} \right).$$

Ex. (ii). $\qquad \dfrac{d^2y}{dx^2} + 6\dfrac{dy}{dx} + 9y = 0.$

$$(a + 3)^2 = 0.$$

$$y = e^{-3x}(Ax + B).$$

68. Illustrations of preceding equations. Simple Harmonic Motion.

Simple Harmonic Motion is defined as the motion performed by a body, moving in a straight line under a force which is directed towards a fixed point in that line and proportional to the distance from it.

Taking the fixed point in the line as origin, the equation giving the distance of the point from the origin at time t is evidently $\dfrac{d^2x}{dt^2} = -kx$, the negative sign being prefixed since the acceleration is inward, i.e. opposes the increase of x.

We have seen that the general solution is

$$x = A\cos\sqrt{k}\,t + B\sin\sqrt{k}\,t.$$

Using a transformation given in Ch. III we may write this

$$x = (A^2 + B^2)^{\frac{1}{2}}\left\{\cos\left[\sqrt{k}\,t - \tan^{-1}\frac{B}{A}\right]\right\},$$

x has therefore a maximum value $\sqrt{A^2 + B^2}$ and a minimum value $-\sqrt{A^2 + B^2}$ and the whole motion takes place between these points. The quantity $\sqrt{A^2 + B^2}$ is called the Amplitude of the motion. Further, we can see that the values of x are repeated periodically, for

$$\cos\{\sqrt{k}\,t - \text{const.}\}$$

repeats its values when t has increased by $\dfrac{2\pi}{\sqrt{k}}$. The motion is therefore periodic and its period is $\dfrac{2\pi}{\sqrt{k}}$.

Initial conditions are usually given to determine A and B. For instance in the usual case of S.H.M. the moving point starts with zero velocity at zero time at $x = a$.

We thus get
$$\left. \begin{array}{l} x = a \text{ when } t = 0 \\ \dfrac{dx}{dt} = 0 \text{ when } t = 0 \end{array} \right\},$$
$$\therefore A = a, \quad B = 0,$$

and the equation reduces to the simple form $x = a \cos \sqrt{k}\, t$. This is clearly the motion performed by the foot of the perpendicular from a point on a circle to the x-axis, when the circle is described with constant angular velocity $\sqrt{k}$. This is the standpoint from which S.H.M. was first considered, and the student will find this treatment of the subject in elementary works on Dynamics.

69. General Oscillatory Motion. The Pendulum.

We notice, in fact, that any equation of the form $\dfrac{d^2\phi}{dt^2} = -\lambda\phi$, where ϕ is some parameter which fixes the position of the body at time t, represents a motion, the phases of which repeat themselves after an interval $\dfrac{2\pi}{\sqrt{\lambda}}$, or, as we say, represents an **Oscillation of period** $\dfrac{2\pi}{\sqrt{\lambda}}$.

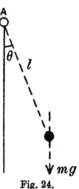

Fig. 24.

Consider, for example, the case of a pendulum formed by a small bob of mass m, at the end of a string of length l, oscillating through a small angle.

The position of the bob is thus determined entirely by the angle θ.

Writing down the equation of motion, as in the preceding chapter,

$$I \frac{d^2\theta}{dt^2} = L;$$

i.e. we have

$$ml^2 \frac{d^2\theta}{dt^2} = -mgl \sin \theta,$$

i.e.

$$\frac{d^2\theta}{dt^2} = -\frac{g}{l}\,\theta,$$

since θ is a small angle.

Therefore we thus see at once that the motion is periodic and its total period $2\pi \sqrt{\dfrac{l}{g}}$. We usually put this result in the form

Time of Vibration (half oscillation) $= \pi \sqrt{\dfrac{l}{g}}.$

The student of Physics will be familiar with many experiments in which a formula of this type is used to determine g *.

70. Resisted Motion. Damped Oscillation.

As a further example consider the following.

A body moves in a straight line under a central force proportional to the distance, but its motion is resisted by a force which is always proportional to the velocity. Determine the motion.

The equation will clearly be

$$\frac{d^2x}{dt^2} = -2\lambda \frac{dx}{dt} - \mu x$$

(negative signs since both forces are retardations), i.e.

$$\frac{d^2x}{dt^2} + 2\lambda \frac{dx}{dt} + \mu x = 0.$$

---

* These formulæ are usually slightly more complicated, for I is usually the M.I. of a compound body, such as a sphere suspended by a string. Again the principle is more important than the formula.

Solving the auxiliary equation $\alpha^2 + 2\lambda\alpha + \mu = 0$ we get as the general solution

$$x = Ae^{(-\lambda + \sqrt{\lambda^2 - \mu})t} + Be^{(-\lambda - \sqrt{\lambda^2 - \mu})t}$$
$$= e^{-\lambda t}(Ae^{\sqrt{\lambda^2 - \mu}\,t} + Be^{-\sqrt{\lambda^2 - \mu}\,t}).$$

If, as is usually the case, $\lambda^2 < \mu$ (the resistance-constant usually being small), the radical $\sqrt{\lambda^2 - \mu}$ is imaginary, and following a previous article we have to write (§ 66)

$$x = e^{-\lambda t}(A \cos \sqrt{\mu - \lambda^2}.t + B \sin \sqrt{\mu - \lambda^2}.t)$$
$$= \sqrt{A^2 + B^2}\,e^{-\lambda t}\left[\cos\left(\sqrt{\mu - \lambda^2}.t - \tan^{-1}\frac{B}{A}\right)\right].$$

Clearly both x and $\dfrac{dx}{dt}$ contain periodic functions, and maximum and minimum values, both of displacement and velocity, occur at intervals of $\dfrac{2\pi}{\sqrt{\mu - \lambda^2}}$.

The motion is semi-oscillatory in character but it possesses the continually decreasing amplitude $\sqrt{A^2 + B^2}\,e^{-\lambda t}$.

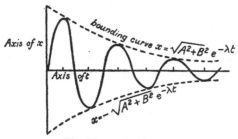

Fig. 25. $x = Ae^{-\lambda t} \cos \mu t$.

Such a motion is known as a Damped Oscillation, and occurs for instance in connection with the oscillatory discharges of Spark Wireless. It can be easily understood diagrammatically as in Fig. 25.

If however $\lambda^2 > \mu$, the solution will not contain trigonometrical functions, but exponentials, and the motion is not oscillatory at all.

71. Equations in which the right-hand side is not zero.

The discussion of the equations in which the right-hand side is not zero, but a function of x, lies beyond the scope of this work. We can however briefly mention one general principle which will often provide the solution of such equations.

Consider the equation

$$\frac{d^2y}{dx^2} + 4\frac{dy}{dx} + 3y = 12 + 9x.$$

We see by inspection that $y = 3x$ is one solution.

Let $y = 3x + v$. Then by substitution

$$\frac{d^2v}{dx^2} + 4\frac{dv}{dx} + 3v = 0,$$

or v is a solution of the same equation with zero on the right-hand side.

We see thus that the solution of any such equation consists of a particular integral together with the Complementary Function (i.e. the general solution formed by writing zero on the right-hand side).

The complete solution of the example given is thus

$$y = Ae^{-3x} + Be^{-x} + 3x.$$

The determination of these particular integrals often demands great ingenuity, and for various methods of doing this the reader is referred to any standard work on Differential Equations.

72. The Equation of Resonance.

We proceed to solve, however, one equation of this type which illustrates an important physical principle.

Consider the equation

$$\frac{d^2x}{dt^2} = -\mu x + L \cos pt.$$

(The equation represents the state of a body acted on by (1) an ordinary central force, and (2) a fixed periodic force.)

We suspect that $A \cos pt$ is a particular solution for some value of A and substituting we get $A = \dfrac{L}{\mu - p^2}$.

Thus $\dfrac{L \cos pt}{\mu - p^2}$ is a particular integral.

The solution is now

$$x = A \cos \sqrt{\mu}\, t + B \sin \sqrt{\mu}\, t + \frac{L}{\mu - p^2} \cos pt.$$

The motion is composed of two Simple Harmonic Motions, one of period $\dfrac{2\pi}{\sqrt{\mu}}$, the other of period $\dfrac{2\pi}{p}$. But it is clear that when μ and p^2 are nearly equal the last member $\dfrac{L}{\mu - p^2} \cos pt$ can become very large even though the actual value of L is small.

It is thus made clear that the superposition of two Simple Harmonic Motions (or two periodic oscillations) may produce an effect far out of proportion to the actual amplitude of either of the motions taken separately.

Thus soldiers have to break step crossing a bridge, in case the rhythm of their march should be equi-periodic with the natural vibration of the bridge; and the combination of the flute with the human voice sometimes produces a very resonant effect. The principle will also be familiar to the students of Wireless Telegraphy and Sound (Heterodyne Reception and Beats).

There are many special devices used in the solution of particular equations. One of these devices (that of multiplying both sides by dy/dx) has been used in a dynamical example of the preceding chapter and should be referred to again, as it is of use in many cases. (Cf. p. 98.)

EXAMPLES X

Solve the equations:

(1) $\dfrac{dy}{dx} + 4y = 0.$

(2) $\dfrac{dy}{dx} = 4y.$

(3) $\dfrac{dy}{dx} = 4y + 4.$

(4) $\dfrac{dy}{dx} = 4y + a.$ (a constant.)

(5) $\dfrac{d^2y}{dx^2} + 4\dfrac{dy}{dx} = 0.$

(6) $\dfrac{d^2y}{dx^2} + 4\dfrac{dy}{dx} = a.$

(7) The outward velocity of a particle moving in a line is 4 times its distance from a point in that line, diminished by 6. If the particle starts from unit distance at zero time, solve the equation of motion and represent the motion graphically.

(8) Solve the equation given by conditions similar to (7) with "inward" velocity substituted for "outward."

Solve the equations:

(9) $\dfrac{d^2y}{dx^2} + 5\dfrac{dy}{dx} + 6y = 0.$

(10) $\dfrac{d^2y}{dx^2} + 5\dfrac{dy}{dx} + 6y = 12.$

(11) $\dfrac{d^2y}{dx^2} - 6\dfrac{dy}{dx} + 9y = 0.$

(12) $\dfrac{d^2y}{dx^2} - 6\dfrac{dy}{dx} + 9y = e^x.$

(13) $\dfrac{d^2y}{dx^2} + 4y = 0.$

(14) $\dfrac{d^2y}{dx^2} - 4\dfrac{dy}{dx} + y = 0.$

(15) $\dfrac{d^3y}{dx^3} + 6\dfrac{d^2y}{dx^2} + 11\dfrac{dy}{dx} + 6y = 0.$

(16) $\dfrac{d^3y}{dx^3} - 3\dfrac{dy}{dx} + 2y = 4x - 6$.

(17) $\dfrac{d^4y}{dx^4} - 16y = 0$.

(18) An elastic string (mod λ) of length l has a bob of mass m attached to the end. Show that if the bob be set in motion by stretching the string it will perform a Simple Harmonic Motion, and find the period.

(19) Explain how the velocity of a bullet may be found by firing it into the bob of a known pendulum.

(20) Find the period of a pendulum composed of a heavy rod of length $2a$ suspended from one end.

(21) Find the period of the pendulum formed by a heavy sphere of radius r attached to a weightless string of length l.

(22) Integrate the equation $\dfrac{d^2s}{dt^2} = f$ to get the usual formulæ of accelerated motion.

(23) Show that for any law of air-resistance of the form Av^n (v is velocity at time t) a falling body will have a maximum or terminal velocity.

(24) Integrate the equation of motion of a falling body when the resistance varies as the velocity.

Find the time taken to reach the terminal velocity if the body falls from rest.

ANSWERS TO EXAMPLES

Examples marked with an asterisk are taken from Higher Certificate Papers by permission of the Board.

(Arbitrary constants to be understood in all cases.)

I A

(1) $\frac{2}{3}x^{\frac{3}{2}}$; $2x^{\frac{1}{2}}$; $\frac{x}{3}+\frac{x^2}{6}+\frac{x^3}{9}$.   (2) $\frac{3}{7}x^{\frac{7}{3}}$; $-\frac{1}{2}x^{-5}$; $-\frac{5}{2}x^{-\frac{2}{5}}$.

(3) $\frac{1}{4}(3x-4)^{\frac{4}{3}}$; $\frac{2}{9}(3x-4)^{\frac{3}{2}}$; $\frac{2}{3}(3x-4)^{\frac{1}{2}}$.

(4) $\log(a+x)$; $\log\dfrac{a+x}{a-x}$; $\dfrac{1}{(n-1)(a-x)^{n-1}}$.

(5) $\log(x+1)$; $x+2\log(x+1)$; $\dfrac{1}{a}\log(ax+b)$;

     $x+\dfrac{c-b}{a}\log(ax+b)$; $\dfrac{A}{a}x+\dfrac{Ba-bA}{a^2}\log(ax+b)$.

(6) $2x^{\frac{1}{2}}+\frac{2}{3}x^{\frac{3}{2}}$; $\frac{1}{2}x^2+\log x$; $\frac{2}{3}\left[(x+1)^{\frac{3}{2}}-x^{\frac{3}{2}}\right]$; $\frac{2}{5}(x+1)^{\frac{5}{2}}-\frac{2}{3}(x+1)^{\frac{3}{2}}$.

(8) $\log(x+\sqrt{x^2-a^2})$.

I B

(1) $\sin^2 x$; $\dfrac{x}{2}+\dfrac{\sin 2x}{4}$; $\dfrac{1}{2}\left[\dfrac{\sin(m+n)x}{m+n}+\dfrac{\sin(m-n)x}{m-n}\right]$;

     $\dfrac{1}{2}\left[\dfrac{\sin(m-n)x}{m-n}-\dfrac{\sin(m+n)x}{m+n}\right]$.

(2) $\log\sec x$; $\sec x$; $\log\sin x-\frac{1}{2}\operatorname{cosec}^2 x$; $\log\tan x$.

(3) $\log\log x$; $\log(e^x+2)$; $\log(e^x-e^{-x})$; $\log(\log\sin x)$.

(4) $\log(x^2-3x+2)$; $x+\log(2x-1)$; $\frac{1}{4}(x^2+2x+2)^2$;

     $\dfrac{x^2}{2}+x+\log(x+1)$; $\frac{1}{2}\log(x^2+2x+2)$.

(5) $\frac{1}{3}\tan^{-1}\dfrac{x}{3}$; $x-2\tan^{-1}\dfrac{x}{2}$; $\tan^{-1}x^2$; $\dfrac{1}{\sqrt{21}}\tan^{-1}\dfrac{x\sqrt{3}}{\sqrt{7}}$.

(6) $-\frac{1}{4}\cos^4 x$; $\frac{1}{3}\sin^3 x-\frac{1}{5}\sin^5 x$; $\sin x-\frac{1}{3}\sin^3 x$.

II

(1) $-\frac{2}{3}(1-x)^{\frac{3}{2}}$.

(2) $\frac{3}{8}(1+2x)^{\frac{4}{3}}$.

(3) $\tan^{-1}x^2$.

(4) $\tan^{-1}x^3$.

(5) $\frac{1}{\sqrt{2}}\tan^{-1}\frac{x+1}{\sqrt{2}}$.

(6) $\sin^{-1}\frac{x+1}{3}$.

(7) $\sec^{-1}(x-1)$.

(8) $\sin^{-1}(2x-1)$ $(z=x-\frac{1}{2})$.

(9) $\tan^{-1}e^x$ (get rid of negative powers).

(10) $\sin e^x$.

(11) a.

(12) $\frac{2}{9}a^{\frac{9}{2}}(2\sqrt{2}-1)$.

(16) $\tan\frac{x}{2}$.

(17) $-\frac{1}{a}\sqrt{\dfrac{2a-x}{x}}$. $\left(\text{Put } y=\dfrac{1}{x}.\right)$

(18) $\frac{1}{a}\sec^{-1}\frac{x}{a}$.

(19) $\sin^{-1}\frac{x-a}{a}$.

(20) $\frac{1}{2}\tan^2 x$.

(21) $\frac{\pi}{32}$.

III

(1) $-(x+1)e^{-x}$; $e^{2x}\left(\dfrac{x}{2}-\dfrac{1}{4}\right)$; $e^x(x^2-2x+2)$;

$e^x(x^3-3x^2+6x-6)$.

(2) $\sin x-x\cos x$; $(x^2-2)\sin x+2x\cos x$;

$(6x-x^3)\cos x+3(x^2-2)\sin x$; $\dfrac{x^2+x\sin 2x}{4}+\dfrac{\cos 2x}{8}$;

$\dfrac{\sin 2x}{8}-\dfrac{x\cos 2x}{4}$.

(3) $\dfrac{x^2}{2}\log x-\dfrac{x^2}{4}$; $\dfrac{x^3}{3}\log x-\dfrac{x^3}{9}$; $\dfrac{x^{n+1}}{n+1}\log x-\dfrac{x^{n+1}}{(n+1)^2}$.

(4) $\frac{1}{2}e^x(\sin x+\cos x)$; $\dfrac{1}{\sqrt{5}}e^x\cos(2x-\tan^{-1}2)$;

$\dfrac{1}{\sqrt{5}}e^x\sin(2x-\tan^{-1}2)$; $e^x\sin x$.

(5) $x\sin^{-1}x+\sqrt{1-x^2}$; $x\tan^{-1}x-\frac{1}{2}\log(1+x^2)$;

$\dfrac{x^2}{2}\sin^{-1}x+\frac{1}{4}x\sqrt{1-x^2}-\frac{1}{4}\sin^{-1}x$; $\dfrac{x^2+1}{2}\tan^{-1}x-\dfrac{x}{2}$.

(6) $\dfrac{x\sqrt{x^2-a^2}}{2}-\dfrac{a^2}{2}\log\dfrac{x+\sqrt{x^2-a^2}}{a}$; $\dfrac{x\sqrt{a^2-x^2}}{2}+\dfrac{a^2}{2}\sin^{-1}\dfrac{x}{a}$.

(7) $\dfrac{e^{3x}}{5}\{\cos(4x-\tan^{-1}\tfrac{4}{3})\}$;

$$e^{5x}\left\{\dfrac{1}{5\sqrt{5}}\sin(10x-\tan^{-1}2)+\dfrac{1}{\sqrt{41}}\sin(4x-\tan^{-1}\tfrac{4}{5})\right\};$$

$$e^{x}\left\{\dfrac{1}{\sqrt{5}}\sin(2x-\tan^{-1}2)+\dfrac{1}{\sqrt{17}}\sin(4x-\tan^{-1}4)\right.$$
$$\left.-\dfrac{1}{\sqrt{37}}\sin(6x-\tan^{-1}6)\right\}.$$

(8) $x\tan x+\log\cos x$; $\dfrac{x}{\sqrt{1-x^2}}\sin^{-1}x+\log\sqrt{1-x^2}$.

(Put $y=\sin^{-1}x$ in last example.)

IV

(1) $\frac{1}{2}\log\dfrac{x+1}{x+3}$.   (2) $\frac{1}{2}\log(x^2+2x+3)-\dfrac{1}{\sqrt{2}}\tan^{-1}\dfrac{x+1}{\sqrt{2}}$.

(3) $\dfrac{1}{25}\left[\log(4-5x)+\dfrac{4}{4-5x}\right]$.   (4) $\log\dfrac{(x-3)^3}{(x-2)^2}$.

(5) $2x-\frac{3}{2}\log(x^2+6x+10)+11\tan^{-1}(x+3)$.

(6) $x-2\log(x^2+2x+2)+3\tan^{-1}(x+1)$.

(7) $\frac{1}{2}\log(x^2-2x-2)-\log(x-1)$.

(8) $\frac{2}{3}\log(x+1)+\frac{1}{6}\log(x^2-x+1)+\dfrac{1}{\sqrt{3}}\tan^{-1}\dfrac{2x-1}{\sqrt{3}}$.

(9) $\log x-\frac{1}{2}\log(x^2+x+1)-\dfrac{1}{\sqrt{3}}\tan^{-1}\dfrac{2x+1}{\sqrt{3}}$.   (10) $1+\dfrac{2\pi}{3\sqrt{3}}$.

(11) $\dfrac{1}{7}\left(\log\dfrac{x-2}{x+2}-\sqrt{3}\tan^{-1}\dfrac{x}{\sqrt{3}}\right)$.

(12) $\dfrac{1}{ab(a^2-b^2)}\left(a\tan^{-1}\dfrac{x}{b}-b\tan^{-1}\dfrac{x}{a}\right)$.

(13) $\dfrac{1}{4\sqrt{2}}\log\dfrac{x^2+\sqrt{2}x+1}{x^2-\sqrt{2}x+1}+\dfrac{1}{2\sqrt{2}}\tan^{-1}\dfrac{x\sqrt{2}}{1-x^2}$.

(14) $\frac{1}{24}\log\dfrac{4+3\sin x}{4-3\sin x}$.

(15) $\frac{1}{4}\log_e\frac{5}{3}$.   (16) $\dfrac{\pi}{2(a+b)}$.

V A

(1) $\frac{1}{3}\cos^3 x - \cos x.$

(2) $-\frac{1}{5}\cos^5 x + \frac{2}{3}\cos^3 x - \cos x.$

(3) $\frac{1}{9}\sin^9 x - \frac{4}{7}\sin^7 x + \frac{6}{5}\sin^5 x - \frac{4}{3}\sin^3 x + \sin x.$

(4) $\frac{1}{3}\sin^3 x - \frac{2}{5}\sin^5 x + \frac{1}{7}\sin^7 x.$

(5) $\sin x + 2\operatorname{cosec} x - \frac{1}{3}\operatorname{cosec}^3 x.$

(6) $\frac{1}{4}\log\dfrac{1+\cos x}{1-\cos x} - \dfrac{1}{2}\dfrac{\cos x}{\sin^2 x}.$

(7) $-\frac{1}{5}\cos^5 x.$

(8) $\frac{1}{5}\cos^5 x - \frac{1}{7}\cos^7 x.$

(9) $\dfrac{1}{4}\log\dfrac{2+\tan\frac{x}{2}}{2-\tan\frac{x}{2}}.$

(10) $\dfrac{1}{4}\log\dfrac{3-\tan\frac{x}{2}}{1-3\tan\frac{x}{2}}.$

(11) $\dfrac{1}{\sqrt{2}}\log\dfrac{1-\sqrt{2}+\tan\frac{x}{2}}{1+\sqrt{2}-\tan\frac{x}{2}}.$

(12) $\log\left(1+\tan\frac{x}{2}\right).$

(13) $\sqrt{2}\tan^{-1}\dfrac{1+\tan\frac{x}{2}}{\sqrt{2}}.$

(14) $\dfrac{1}{\sqrt{a^2+b^2}}\log\dfrac{a\tan\frac{x}{2}-b+\sqrt{a^2+b^2}}{a\tan\frac{x}{2}-b-\sqrt{a^2+b^2}}.$

(17) $\dfrac{\pi-a}{\sin a}.$

(18) (a) $\dfrac{\pi}{2\sqrt{1-k^2}};\ k<1.$ (b) $\dfrac{1}{2\sqrt{k^2-1}}\log\left(1-\dfrac{1}{k^2}\right);\ k>1.$

(19) $\dfrac{1}{\sin a}\log\dfrac{1+\tan\frac{x}{2}\tan\frac{a}{2}}{1-\tan\frac{x}{2}\tan\frac{a}{2}}.$

V B

(1) $\dfrac{3x}{8} - \dfrac{\sin 2x}{4} + \dfrac{\sin 4x}{32}.$

(2) $\dfrac{5x}{16} + \frac{15}{64}\sin 2x + \frac{3}{64}\sin 4x + \frac{1}{192}\sin 6x.$

(3) $\dfrac{x}{8} - \dfrac{\sin 4x}{32}$.

(4) $\dfrac{x}{16} - \dfrac{\sin 2x}{64} - \dfrac{\sin 4x}{64} + \dfrac{\sin 6x}{192}$.

(5) $\tan x - \cot x$.

(6) $-\tfrac{1}{2}(\cot x \operatorname{cosec} x) - \tfrac{1}{2}\log(\cot x + \operatorname{cosec} x)$.　(Put $c = \cot x$.)

(7) $\tfrac{1}{3}\sec^3 x - \tfrac{1}{2}\log\dfrac{1+\cos x}{1-\cos x}$.

(8) $I_n = \dfrac{1}{n+1}\sin^{n+1} x \cos x + \dfrac{n+2}{n+1} I_{n+2}$.

(9) (a) $I_{p,\,q} = \dfrac{1}{p+1}\sin^{p+1} x \cos^{q-1} x + \dfrac{q-1}{p+1} I_{p+2\ q-2}$.

(b) $I_{p,\,q} = \dfrac{1}{q+1}\sin^{p-1} x \cos^{q+1} x + \dfrac{p-1}{q+1} I_{p-2,\,q+2}$.

(10) $\dfrac{35\pi}{256}$, $\dfrac{35\pi}{128}$, $\dfrac{35\pi}{128}$.

(11) $\dfrac{5\pi}{32}$, $\dfrac{16}{35}$, 0, 0.

(12) $\dfrac{3\pi}{256}$, $\dfrac{3\pi}{128}$.

(13) 0.

(15) $\tfrac{1}{2}\tan\phi\sec\phi + \tfrac{1}{2}\log\tan\left(\dfrac{\pi}{4} + \dfrac{\phi}{2}\right)$;

$\tfrac{1}{4}\tan\phi\sec^3\phi + \tfrac{3}{8}\tan\phi\sec\phi + \tfrac{3}{8}\log\tan\left(\dfrac{\pi}{4} + \dfrac{\Phi}{2}\right)$.

(16) $\dfrac{x^6(\log x)^2}{5} - \dfrac{2x^5(\log x)}{5^2} + \dfrac{2x^5}{5^3}$.

(17) $I = -\dfrac{1}{m+p+1}\cdot x^{m+1}(a^2 - x^2)^{\frac{p}{2}}$

$+ \dfrac{pa}{m+p+1}\displaystyle\int x^m (a^2 - x^2)^{\frac{p}{2}-1}\,dx$.

(19) $I_m = -\dfrac{x^{m+\frac{1}{2}}\sqrt{(2a-x)^3}}{m+2} + \dfrac{2m+1}{m+2}\cdot a I_{m-1}$.

(20) $\dfrac{21\pi a^6}{16}$.

VI

(1) (a) $11\tfrac{1}{2}$ units of area.　(b) $15\tfrac{1}{2}$ units of area.

(2) (a) $\tfrac{3}{4} a^{\frac{2}{3}} h^{\frac{4}{3}}$　(b) $\dfrac{h^4}{4a^2} + \dfrac{h^3}{3a}$.　(3) 2 units of area.　(4) $\dfrac{16a^2}{3}$.

(6) The integration of $\sin^2\theta$ on which it depends assumes the formulæ of Radian measure, which are equivalent to the formulæ for area and circumference of a circle.

(7) $\frac{8}{15}$ units of area.   (8) πa^2.   (10) $2:3$.   (11) $\frac{44}{15}$ units of area.

(12) $2a^2\left(1-\dfrac{\pi}{4}\right)$; $2a^2\left(1+\dfrac{\pi}{4}\right)$.   (13) $4a^2$.

VII A

(1) $\dfrac{\pi a^2}{16}$.   (2) $\dfrac{\pi(a^2+b^2)}{2}$.   (3) $\dfrac{a^2}{2}$.   (4) $\pi\left(a^2+\dfrac{b^2}{2}\right)$.

(5) (i) $\dfrac{3\pi a^2}{2}$; (ii) $\left(a^2+\dfrac{b^2}{2}\right)\left(\pi-\cos^{-1}\dfrac{a}{b}\right)$.   (6) $\dfrac{a^2}{4}(\epsilon^{2\pi}-1)$.

(7) $\frac{1}{3}\pi h\,(r_1{}^2+r_1 r_2+r_2{}^2)$.

(8) $\pi\,[a^2\{(a^2-r_1{}^2)^{\frac{1}{2}}-(a^2-r_2{}^2)^{\frac{1}{2}}\}-\frac{1}{3}\{(a^2-r_1{}^2)^{\frac{3}{2}}-(a^2-r_2{}^2)^{\frac{3}{2}}\}]$.

(9) $\frac{4}{3}\pi ab^2$.

(10) The spheroid is "prolate," i.e. of revolution about the minor axis; $\therefore$ its volume is $\frac{4}{3}\pi a^2 b$. Hence the error is $\frac{4}{3}\pi a^2(a-b)$ in excess, i.e. $\left(\dfrac{a-b}{b}\right)100\,°/_\circ=100\left(\dfrac{a}{b}-1\right)°/_\circ$. But $a^2-b^2=a^2.\frac{1}{3600}$; $\therefore \dfrac{a}{b}=(1-\frac{1}{3600})^{-\frac{1}{2}}=1+\frac{1}{7200}$ approx. $\therefore$ error$=\frac{1}{72}°/_\circ$.

(11) $2\pi ah^2$.   (12) $\dfrac{\pi^2 a^3}{2}$.   (13) $\pi c^2\left(r-\dfrac{c}{3}\right)$.   (14) $\dfrac{128\pi r^3}{375}$.

(15) Where $x=\dfrac{144a}{49}$.

VII B

(2) The arc from $\theta=\beta$ to $\theta=a$ is
$$a\left[\frac{\theta\sqrt{\theta^2+1}}{2}+\tfrac{1}{2}\log(\theta+\sqrt{\theta^2+1})\right]_a^\beta.$$

(3) $4a$.   (4) $\dfrac{4b(a+b)}{a}\cos\dfrac{a}{2b}\,\theta_1$.   (5) $2a\log\dfrac{a+a}{a-a}-a$.

(6) $\dfrac{1}{27\sqrt{a}}\left[(4a+9x)^{\frac{3}{2}}\right]_{x_1}^{x_2}$.   (7) $6a$.

VIII

(1) At a distance $\dfrac{4r}{3\pi}$ along the bisecting radius.

(2) At a distance $\dfrac{2r}{\pi}$ from centre.

(3) Along the axis of the hemisphere at a distance $\dfrac{3r}{8}$.

(4) At $\left(\dfrac{4a}{3\pi},\ \dfrac{4b}{3\pi}\right)$. \qquad (7) $\bar{x}=\dfrac{8a}{5m^2}$, $\bar{y}=\dfrac{2a}{m}$.

(8) Taking the planes at distances d_1 and d_2 from the centre the expression for $\bar{x}$ is

$$\frac{d_1+d_2}{2}\cdot\frac{a^2-\frac{1}{2}(d_1{}^2+d_2{}^2)}{a^2-\frac{1}{3}(d_1{}^2+d_1 d_2+d_2{}^2)}.$$

The surface $=\pi a\,(d_1-d_2)$.

(9) On the axis at a distance $\dfrac{h}{4}\cdot\dfrac{r^2+2rR+3R^2}{r^2+rR+R^2}$ from the smaller end.

(10) Surface $=\pi r\sqrt{h^2+r^2}$; c.g. $\frac{2}{3}h$ along axis.

(11) c.g. divides rod in ratio $2:1$. \qquad (12) $\bar{x}=h\cdot\dfrac{m+3n}{2m+4n}$.

(13) $V=2\pi^2 a^2 d$; $S=4\pi^2 ad$. \qquad (14) $V=\pi a^3\sqrt{2}$; $S=4\pi a^2\sqrt{2}$.

(15) $\dfrac{\pi^2 a^3}{12}$. \qquad (17) $V=\cdot32625\pi$ cu. in.; $S=4\cdot36\pi$ sq. ins. approx.

IX A

(1) $\dfrac{Ma^2}{3}$; $\dfrac{Mb^2}{3}$. \qquad (2) $M\dfrac{b^2+c^2}{3}$; $M\dfrac{c^2+a^2}{3}$; $M\dfrac{a^2+b^2}{3}$.

(3) $M\dfrac{r^2}{2}$. \qquad (4) $\dfrac{2Ma^2}{3}$. \qquad (5) $\dfrac{Mb^2}{4}$; $\dfrac{Ma^2}{4}$. \qquad (6) $M\dfrac{r^2}{2}$.

(7) $\dfrac{5Ma^2}{4}$. \qquad (8) $\dfrac{3Ma^2}{2}$. \qquad (9) $M\dfrac{r^2}{3}$ ($2r=$ base of $\triangle$).

(10) $\dfrac{2M}{5}\cdot\dfrac{b^5-a^5}{b^3-a^3}$. \qquad (11) $\dfrac{5^7\pi^2}{36}$ ft. lbs.; $\dfrac{5^6\pi}{288}$ secs.; $\dfrac{5^7\pi}{864}$ revs.

(12) $\dfrac{5^6\pi}{576}$ secs.

(13) k.e. $=\dfrac{3M}{2}\left(\dfrac{d\theta}{dt}\right)^2$; p.e. $=-\dfrac{3M}{2}g\sin\theta$. Angular velocity $=\sqrt{g}$ radians per sec.

IX B

(1) Measure the area under the curve between ordinates corresponding to the given intervals.

(2) $pv \log 3$. (3) 6 foot lbs.

(4) Extension × mean of initial and final tensions.

(5) $\frac{3}{8}$ foot lbs. (6) To $\left(1 + \dfrac{1}{\sqrt{3}}\right)$ feet. (7) $\frac{2}{3}r$.

(8) The angle is such that the moment about the hinge of the resultant pressure acting at the centre of pressure is equal to the moment of the weight of the lid.

X

(1) $y = Ae^{-4t}$. (2) $y = Ae^{4t}$. (3) $y = Ae^{4t} + 1$.

(4) $y = Ae^{4t} + \dfrac{a}{4}$. (5) $y = Ae^{-4t} + B$. (6) $y = Ae^{-4t} + B + \dfrac{ax}{4}$.

(7) $x = \frac{5}{2}e^{4t} + \frac{3}{2}$. (8) $x = -\frac{1}{2}e^{-4t} - \frac{3}{2}$. (9) $y = Ae^{-3x} + Be^{-2x}$.

(10) $y = Ae^{-3x} + Be^{-2x} + 2$. (11) $y = e^{3x}(Ax + B)$.

(12) $y = e^{3x}(Ax + B) + \dfrac{e^x}{4}$. (13) $y = A \cos 2x + B \sin 2x$.

(14) $y = e^{2x}(A \cos \sqrt{3}\,x + B \sin \sqrt{3}\,x)$.

(15) $y = Ae^{-x} + Be^{-2x} + Ce^{-3x}$. (16) $y = e^x(Ax + B) + Ce^{-2x} + 2x$.

(17) $y = Ae^{2x} + Be^{-2x} + C \cos 2x + D \sin 2x$.

(18) Period $= 2\pi \sqrt{\dfrac{lm}{\lambda g}}$. (20) $2\pi \sqrt{\dfrac{4a}{3g}}$.

(21) $2\pi \sqrt{\dfrac{l^2 + 2lr + \frac{7}{5}r^2}{(l + r)g}}$.

(23) For the downward acceleration $= g - \dfrac{Av^n}{m}$, which $= 0$ when $v = \sqrt[n]{\dfrac{mg}{A}}$, which will thus be the max. velocity.

(24) The equation is $\dfrac{d^2x}{dt^2} = -\lambda \dfrac{dx}{dt} + g$ (x is the vertical height at time t) and the terminal velocity $\dfrac{g}{\lambda}$ is only reached after infinite time.

CPSIA information can be obtained
at www.ICGtesting.com
Printed in the USA
LVHW051720220520
656325LV00001B/56

9 781316 612699